Fundamentals of Amorphous Semiconductors

Report of
THE AD HOC COMMITTEE ON THE
FUNDAMENTALS OF AMORPHOUS
SEMICONDUCTORS

NATIONAL MATERIALS ADVISORY BOARD
NATIONAL RESEARCH COUNCIL

NATIONAL ACADEMY OF SCIENCES
WASHINGTON, D.C. 1972

This study by the National Materials Advisory Board was conducted under Contract No. NOOO14-67-A-0244-0022 with the Office of Naval Research.

Members of the National Materials Advisory Board study groups serve as individuals contributing their personal knowledge and judgments and not as representatives of any organization in which they are employed or with which they may be associated.

The quantitative data published in this report are intended only to illustrate the scope and substance of information considered in the study, and should not be used for any other purpose, such as in specifications or in design, unless so stated.

No portion of this report may be republished without prior approval of the National Materials Advisory Board.

Library of Congress Catalog Card Number 75-188496
ISBN 0-309-01944-3

Available from
Printing and Publishing Office
National Academy of Sciences
2101 Constitution Avenue, N.W.
Washington, D.C. 20418

Printed in the United States of America

THE AD HOC COMMITTEE ON THE FUNDAMENTALS OF AMORPHOUS SEMICONDUCTORS

Chairman

HENRY EHRENREICH, Gordon McKay Professor of Applied Physics,
Division of Engineering & Applied Physics,
Harvard University, Pierce Hall, Cambridge, Massachusetts 02138

Members

EVAN J. FELTY, Materials Science Laboratory,
Xerox Corporation, Xerox Square, Rochester, New York 14603
HANS P. R. FREDERIKSE, Chief, Solid State Physics Section,
National Bureau of Standards, Washington, D.C. 20234
B. I. HALPERIN, Theoretical Group,
Bell Telephone Laboratories, Murray Hill, New Jersey 07971
ROLF W. LANDAUER, International Business Machines Corporation,
Thomas J. Watson Research Center,
P.O. Box 218, Yorktown Heights, New York 10598
JAN TAUC, Professor, Division of Engineering,
Brown University, Providence, Rhode Island 02912
DAVID TURNBULL, Gordon McKay Professor of Applied Physics,
Division of Engineering & Applied Physics,
Harvard University, Pierce Hall, Cambridge, Massachusetts 02138

NMAB *Staff*

DONALD G. GROVES, National Materials Advisory Board,
National Research Council,
National Academy of Sciences, National Academy of Engineering,
2101 Constitution Ave., N.W., Washington, D.C. 20418

Liaison Representatives

HARRY FOX, Director, Chemistry Program,
Code 472, Office of Naval Research,
800 North Quincy St., Arlington, Virginia 22217
FRANK B. ISAKSON, Director, Physics Program,
Code 421, Office of Naval Research,
800 North Quincy St., Arlington, Virginia 22217
ROBERT G. MORRIS, General Physicist,
Code 421, Office of Naval Research,
800 North Quincy St., Arlington, Virginia 22217

Acknowledgments

This report has benefited considerably from the help of many persons. The dedication and enthusiasm of the Committee members made this project a thoroughly stimulating and rewarding one. Indeed, the group became sufficiently close-knit during its work that this report realistically can be said to represent a unified point of view shared by all of us, rather than a set of compromises among widely divergent opinions.

The Committee itself has profited greatly from technical presentations and comments from a number of colleagues.

We wish to thank the following for taking the time to participate in one of the meetings to make tutorial presentations:

R. S. Allgaier, U.S. Naval Ordnance Laboratory, White Oak, Silver Spring, Maryland

Brian G. Bagley, Bell Telephone Laboratories, Murray Hill, New Jersey

Marc H. Brodsky, International Business Machines Corp., Yorktown Heights, New York

Alan G. Chynoweth, Bell Telephone Laboratories, Murray Hill, New Jersey

William Doremus, Picatinny Arsenal, Dover, New Jersey

Hellmut Fritzsche, University of Chicago, James Franck Institute, Chicago, Illinois

James Krumhansl, Department of Physics, Cornell University, Ithaca, New York

Dean L. Mitchell, Naval Research Laboratory, Washington, D.C.

Albert Rose, RCA David Sarnoff Research Center, Princeton, New Jersey

P. J. Walsh, Picatinny Arsenal, Dover, New Jersey

We are particularly grateful to S. R. Ovshinsky for his cooperation in supplying information to the Committee and for his courtesy in inviting several members to visit Energy Conversion Devices, Inc., for more extensive discussions with him and his collaborators.

One or another of us has also benefited from personal conversations with the following: D. Adler, P. W. Anderson, A. I. Bienenstock, M. H. Cohen, J. de Neufville, T. M. Donovan, J. D. Dow, E. A. Fagen, J. Feinleib, G. C. Feth, V. Heine, H. K. Henisch, T. W. Hickmott, S. C. Moss, M. B. Myers, R. G. Neale, W. Paul, J. C. Phillips, D. Redfield, W. E. Spicer, A. H. Sporer, M. D. Tabak, J. Trommel, K. Weiser, and R. Zallen.

Dr. Robert G. Morris, one of the liaison representatives of the Office of Naval Research, has been particularly helpful during the course of our meetings in making a number of useful comments based on his own wide knowledge of the amorphous semiconductor field, and by providing the Committee with a set of notes concerning its proceedings which have been of great aid in preparing this report.

The Committee is very grateful to Don Groves of the staff of the National Materials Advisory Board for providing the conditions that have permitted it to function effectively, and for a number of valuable perspective comments, all of which have made the completion of this task far easier than it would otherwise have been.

Finally, I want to acknowledge the considerable help provided by Miss Linda Freeman, my secretary at Harvard, in assembling the final version of the manuscript.

HENRY EHRENREICH, *Chairman*
National Materials Advisory Board
Ad Hoc Committee on the
Fundamentals of Amorphous Semiconductors

Abstract

The study of glasses has been important historically because of their great technological usefulness. One class of these materials, amorphous semiconductors, has evoked a great deal of interest during the past few years. This interest stems in part from the fact that solid state physics, after attaining a remarkably high level of scientific understanding of crystals, can now hope for comparable achievements in connection with disordered materials. Of equal importance is the fact that the metastability of amorphous semiconductors provides them with certain unique properties that may be of considerable technological significance.

This report is intended to provide an overview of the field, its present standing, and its promise. The fundamental structural and electronic properties and the present level of understanding of these properties is of primary concern. However, much of the progress in solid state physics has traditionally been motivated by technological considerations. Therefore, the principal aspects of the physics underlying the more important amorphous semiconductor devices are discussed, as well as the technological setting in which this new field finds itself.

Preface

As a result of increased scientific activity and interest in the field of amorphous materials, the Office of Naval Research agreed that the National Materials Advisory Board (NMAB) of the National Research Council initiate an appropriate committee study that would address this challenging subject in a broad fundamental manner. In this way, it was hoped that the study might be helpful in organizing the presently available information about the field, assessing its importance to physics and materials research, and providing a perspective setting for future investigations.

In this assignment, which was accepted in October 1970, NMAB was requested to study those areas of materials science and solid state physics that are pertinent to:

a. the unique physical properties of amorphous materials,

b. a characterization of amorphous materials and the relation of physical properties to the characterization parameters,

c. a description of the fundamental properties of amorphous materials,

d. fruitful theoretical analyses of the disordered state, and

e. a discussion of the physics underlying amorphous semiconductor devices.

Based on its findings, the Committee was requested to document in its final report "promising areas of research appropriate to the opportunities and problems and to suggest in this report the kind and extent of research necessary to advance amorphous materials science and technology."

Because of the nature of this request, the Committee recognized that an adequate discussion would involve placing the subject in its proper technological setting. It was felt that this could be accomplished without making definitive assessments of comparative technologies. The tutorial discussion given in Section VII should, however, provide some perspective.

In order to circumscribe the scope of the report, the Office of Naval Research suggested that the amorphous materials under the Committee's purview should include primarily elements and mixtures of elements from Columns IV, V, and VI of the periodic table and that materials like gases, liquids, plastics, liquid crystals, organic materials, structural materials such as amorphous alloys and carbon, and both silicate and oxide glasses would be excluded from extensive consideration. This delimitation of materials appeared appropriate to the Committee, since the semiconducting glasses to be emphasized here are just those responsible for the great interest in this field that has developed during the recent past. As already pointed out, the oxide glasses are technologically very significant. Electronic phenomena in amorphous oxide films are the subject of an extensive recent review (Dea 71).* A recent study of interest concerning the physics of amorphous materials in general was initiated by the British Science Research Council (Sr 70).

Since it is the intent here to present a reasonably concise overview of this field with regard to its present standing, its promise, and existing needs and opportunities for further research, this report obviously cannot present an encyclopedic survey even of the amorphous materials that remain after the oxide and metallic glasses have been eliminated. Accordingly, only those amorphous semiconductors of greatest fundamental and/or technological interest will be considered in any detail. These are the elemental glasses, Se, Ge, and Si and the chalcogenide mixtures As–S, As–Se, Te–Ge, and Te–As–Si. The last three have well known applications respectively in xerography and memory and threshold switching.

---

* The convention for referencing employed in this report is explained at the beginning of Section IX which lists the literature citations. Wherever possible throughout the report, the references are to review articles rather than the original papers. Such articles are marked R in the reference section.

The NMAB Ad Hoc Committee on the Fundamentals of Amorphous Semiconductors was formed in October 1970 and conducted its first meeting in November (9–10) 1970. The full Committee held six formal two-day meetings during the period November 1970 through June 1971. In addition, there were several smaller sessions involving various groups of Committee members.

Contents

I

Introduction

It is customary to restrict the designation "glass" to those amorphous solids that have been formed by cooling a liquid. However, it is doubtful that "glasses" so defined differ sharply in microscopic character from amorphous solids with the same composition formed in other ways. Thus the terms "amorphous solids" and "glasses" will be taken to be equivalent in this report. Glasses can be metallic, semiconducting, or insulating. The forces bonding the atoms are analogous to those found in crystals. The chemical bonding can be covalent, ionic, metallic, van der Waals, or hydrogen bonding, or combinations of these. Most glasses, however, fall into the predominantly covalent category. Because of their metastability, glasses exhibit properties that are quite unique and remarkable. They do not undergo a first-order-phase transition at the melting temperature. Instead, they soften gradually at sufficiently high temperatures and pass more or less continuously into the liquid state. The molten glass may either return to its original state if it is cooled sufficiently rapidly, or crystallize if it is cooled slowly. Glasses containing several constituents may exhibit a separation into phases having different compositions on a very minute spatial scale. These structural transformations have a qualitative influence on the

1

electrical and optical properties in various types of glasses. These are of interest not only as phenomena in themselves but also because of their technological significance. Finally, the mere fact that glasses are structurally disordered suggests that their properties can be relatively insensitive to high-energy radiation and bombardment.

The oxide glasses are, perhaps, the most familiar. The soda-lime-silicate glasses (mixtures of Na_2O, CaO, and SiO_2) are good dielectrics, thermal insulators, and optical transmitters. Because they soften gradually with increasing temperature, it is possible to pour, mold, roll, press, and float ordinary window glass, processes that are essential in its manufacture. Many, though not all, oxide glasses are insulators with conductivities less than 10^{-8} Ω^{-1} cm^{-1}. This fact, as well as the natural tendency of metals to oxidize, makes these materials very useful in solid state device technology.

Semiconducting glasses (or vitreous semiconductors) were not investigated to any large degree before 1955. In contrast to the insulating glasses, the conductivity in these substances is electronic rather than ionic. As a result, the conductivity is larger, ranging from 10^{-13} to 10^{-3} Ω^{-1} cm^{-1}. While some of the semiconducting glasses are oxides, the most widely studied examples do not contain oxygen. Instead they contain another constituent, such as S, Se, Te of group six of the periodic table. Such elements are called "chalcogens" and the glasses involving them are known as chalcogenides. The chemical bonding in such glasses is predominantly covalent with a smaller ionic contribution, although cases involving mixed covalent and van der Waals binding are also frequently encountered in materials such as Se.

The chalcogenide glasses have received a great deal of attention because of their established or possible importance in connection with electrophotography, infrared transmitting windows, electronic switching, and electronic and optical memory applications. Work at Energy Conversion Devices, Inc. (ECD), has particularly spurred the development of applications for chalcogenide glasses.

Elemental amorphous Se has been investigated extensively in part because it forms the essential ingredient of the photosensor involved in xerography. In practice the commercial compositions may contain some As and traces of other elements. As–Se glasses have also been studied at RCA in connection with vidicon applications. Indeed, the fundamental properties of these glasses have received considerable attention both in this country and the Soviet Union (Ko 64). The interest in technological applications of chalcogenide glasses has stimulated interest in other chalcogenide glass compositions, such as those belonging to the Ge–Te family.

Crystalline Si and Ge are among the best understood solids. Their amorphous forms are of interest particularly in connection with fundamental research directed toward exploring physical differences between the crystalline and amorphous states. The delineation of these differences would be expected to be simpler in elemental glasses that contain only structural and not compositional disorder.

The metallic glasses usually occur as compounds of the form A_3B to A_5B, where A is a noble or transition metal and B is a metalloid like Si, Ge, or P. While they exhibit a variety of interesting properties, including radiation hardness, they have not yet found significant use in electronic technology. Since amorphous semiconductors have been observed to crystallize in the neighborhood of conventional metallic contacts, speculation has focused on the possibility of using amorphous metal contacts on semiconducting glasses in order to prevent this from happening.

It should be emphasized that the study of glasses is important, not only for technological reasons but also, more fundamentally, because they are systems having structural and possibly compositional disorder. Until very recently solid state physics has been concerned almost exclusively with crystalline materials. Considerations of disorder emphasized effects arising, for example, from lattice vibrations, point defects, and impurities and dislocations in small concentrations that only influence the crystalline properties weakly. However, during recent years, emphasis has been increasingly given to the investigation of the properties of strongly disordered materials, such as liquids, binary substitutional alloys, and amorphous materials. Clearly, an increased understanding of liquids and alloys will be of benefit to those investigating amorphous semiconductors, just as further theoretical insight concerning the materials considered in this report will aid those investigating liquids and alloys.

Amorphous–Crystalline Transformations

The physical, chemical, and mechanical properties of amorphous materials can all be strongly affected by the transformation to the crystalline state. The changes in electrical and optical properties have already been noted. Some examples, representative of the extent of these changes, may be useful. The room temperature resistivity of amorphous Ge and Si films can be as much as five orders of magnitude larger than that of the corresponding polycrystalline films. The extent of the change depends sensitively on the details of the film preparation. By contrast in As_2S_3 and As_2Se_3, the glasses are less resistive than the corresponding crystals. The electrical band gap, as determined from the temperature dependence of the conductivity, is respectively 0.2 and 0.55 eV in c-InSb and a-InSb.* In c- and a-Te, the corresponding quantities are 0.33 and 0.87 eV. The index of refraction in Te and Se decreases respectively by 40 percent and 12 percent in going from the crystalline to the amorphous state. On the other hand, in Ge the index changes but slightly in the opposite direction. The foregoing results are representative of the simplest types of measurements. As will be seen later, even the Hall effect and thermoelectric power, whose measurements represent no difficulty in many crystalline semiconductors, are still poorly established quantitatively in amorphous materials. Further citation of experimental results in the present context might therefore be misleading.

There have been qualitative observations of changes in chemical properties such as wettability, reactivity, adhesion, and solubility resulting from amorphous–crystalline transformations. Mechanical properties, such as hardness, thermal expansion, and sound velocity are similarly affected (Ov 71c).

This unique potential for change in amorphous materials is of both fundamental and technological importance. Its measurement and interpretation is a challenging and important problem for the solid state physicist and chemist; its exploitation is a challenge for the ingenious inventor. Proposals have been made to utilize these and other effects for optical mass memories, memory and threshold switches, electroluminescent displays, nonimpact lithographic plates, and imaging applications including photography and copiers (Ov 71c).

As an example of how these unique properties can be utilized technologically, the memory device takes explicit advantage of the fact that glasses are energetically metastable. In the Te-based glasses, the Te-rich phases tend to segregate from the rest at sufficiently high temperatures. Such temperatures can be achieved by joule heating. Phase separation can also be achieved by photocrystallization. Depending on the maximum temperature and the rate of cooling, the glass then settles either into a state containing crystalline filaments or returns to its initial amorphous state. The two differ by orders of magnitude in conductivity. The mechanisms for memory switching will be discussed in Section VI.

---

* In this report, the prefixes c- and a- will be used to specify the crystalline and amorphous states of the given material unless it is clear otherwise which is being referred to.

OUTLINE

In approaching this report, the reader should bear in mind that the field of amorphous materials is a rapidly developing one. Some of the questions asked here may well have been answered by the time this document appears, and others, perhaps not even alluded to, may have taken their place. Neither should the reader expect a comprehensive survey of the entire area, for this is not meant to be a review in the sense the term is generally understood by the scientific community. It can and should be viewed as a broad survey addressed to those interested in learning about the field, as well as the existing incentives for pursuing investigations directed toward either fundamental or technological ends. Not everyone may necessarily be interested in all sections of this report. The following outline may help the reader to find the information he seeks. While the Committee has assembled a fairly copious set of references, this list is not meant to be in any sense complete. It is, nevertheless, hoped that it will suffice to serve as an entry to various aspects of the literature.

Section II is concerned with the fundamental ideas, already touched upon here, that underlie the structure and the thermodynamic properties of glasses. Such basic concepts as the "ideal" glass, the metastability of the amorphous state, the parameters such as the glass temperature which characterize a given material, and the bonding forces are given detailed attention. These ideas are of central importance because they lead to an understanding of effects, like phase separability and photocrystallization that are unique to the amorphous state.

The third section presents an overview of some of the basic methods used to prepare amorphous materials in either film or bulk form. The fourth section deals with experimental tools that should be useful in characterizing a glass. The delineation of parameters that must be measured, in order to specify a given sample sufficiently uniquely that it can be duplicated either at the same laboratory or elsewhere, is of great importance. Unfortunately, however, amorphous materials as prepared in the laboratory are sufficiently complicated that it is impossible to specify a set of such parameters completely at the present time.

The fifth and longest section of the report is concerned with the fundamental properties of some of the extensively studied amorphous semiconductors, the interpretation of these properties in terms of simple physical models, as well as some of the basic approaches that may be useful in the development of *ab initio* theories. This section, which will be of principal interest to those engaged in basic research, consists of several parts dealing, respectively, with optical, electrical, magnetic, and lattice properties, physical models, and basic theories. Of these, the first four emphasize the experimental aspects of the subject. An attempt is made to appraise the present state of knowledge in terms of opportunities available for further research. However, it must be realized that because many experiments have not yet been done sufficiently reproducibly on well-characterized samples, such an assessment of opportunities is frequently based on rather scant information.

The remaining parts of this report are addressed to the more device-oriented scientists. Section VI is concerned with device physics and presents an overview of the phenomenology used to describe some of the more familiar applications of amorphous materials, such as the threshold switch, and the electrical and optical memories. For completeness, some perspective comments concerning the electronic applications of oxide glasses are made here. Section VII attempts to survey the technological setting in which the amorphous semiconductor technology finds itself. In particular, it discusses some of the other available or suggested technologies, whose products perform functions similar to those seen as promising in amorphous semiconductor devices. Very few, if any, attempts will be made to provide a relative assessment since this can be done meaningfully only with access to proprietary information, and because it is not the Committee's function to provide such an appraisal. A broad-brush survey is included here to present as complete a picture as possible of the amorphous semiconductor field.

Section VIII contains summary and perspective statements as well as a variety of recommendations. Some of these concern further research in various areas, which, in the opinion of the Committee, might profitably be pursued in the future. These recommendations should be viewed in the context of the discussion contained in the body of the report. Section IX is devoted to literature references.

II

Structure and Bonding in Amorphous Solids

Macroscopically, the amorphous solid is distinguished from the fluid by its high resistance to shear deformation, i.e., by its relatively high-shear viscosity. Practically, we consider a body solid (Con 54) when its shear viscosity, η, exceeds 10^{15} poise though the "glass temperature," T_g, is taken to be the temperature at which $\eta = 10^{13}$ poise. It is often found that the time constant, τ, for changes in molecular configuration within an amorphous system, scales roughly as the shear viscosity. According to this scaling law, τ should be of the order of 20 minutes at the glass temperature and one day at $\eta = 10^{15}$ poise.

Microscopically, the basic distinction between solids and fluids might be made in terms of the nature of the molecular motions (Tu 69b); a substantial fraction of these motions is translational or "diffusive" in a fluid, while in a solid, whether amorphous or crystalline, the motions are almost wholly oscillatory. Thus, in contrast with the fluid, in the solid there exists a well-defined set of positions about which the molecules oscillate. These positions are characterized by translational symmetry in the crystal, but in an amorphous solid their pattern is aperiodic. In the crystal, interpositional changes of molecules occur without alteration of the position pattern. Such interpositional exchanges do alter the pattern in an amorphous material near its glass temperature, but it is possible that they would not change the pattern in the hypothetical "ideal" glass.

It has not been proven theoretically that the state of minimum energy of any substance is crystalline rather than amorphous. However, experiment has shown that nearly all pure substances are more stable in some crystalline than in an amorphous solid form. It has been pointed out (Tu 69b) that this generalization may not hold for some systems that are *constrained* to be compositionally disordered. However, it follows that to form an amorphous solid, the ordering processes (crystallization in the case of simple pure substances; compositional ordering followed by crystallization for some mixtures) that are favored thermodynamically must somehow be bypassed. This might be accomplished, for example (Tu 69a), by cooling the liquid at a sufficiently rapid rate or by various deposition (vapor, electro, etc.) techniques.

When glass is formed by cooling a liquid, it is often observed (Kau 48) that the heat capacity and thermal expansivity drop sharply in the vicinity of T_g, as defined above. However, the temperature at which these abrupt changes occur is lower for lower cooling rates, and it simply marks the point below which the amorphous system is no longer in internal configurational equilibrium. That this equilibrium is not achieved in the glasses of ordinary experience is to be expected in view of the very large time constants noted earlier for configurational changes at $T < T_g$. Also, it is not likely that amorphous solids formed by the various deposition techniques are in internal equilibrium. It follows that, at the same temperature and pressure, two amorphous solid specimens with the same compositions may still differ somewhat in internal structure. This behavior is quite analogous to that of a compositionally disordered crystalline alloy at temperatures where the time constant for interpositional exchanges is very long. An amorphous solid, if constrained from crystallizing, would presumably relax after an infinite time to an "ideal" amorphous state of minimum enthalpy and entropy (Gi 58, Coh 60, 64). The structural characteristics of ideal glasses are considered later.

If we classify condensed materials according to the type of bonding responsible for their coherence, i.e., covalent, metallic, ionic, van der Waals, or hydrogen, every class contains some members that can be put into the amorphous solid form (Tu 69a). In general, the tendency to amorphous solid formation is greatest in some covalently bonded materials, and least in most ionic and metallically bonded materials.

The problem of whether the structure of amorphous solids is, in general, distinct and unique, or only trivially different from that of a crystalline solid, has persisted for a long while without being resolved definitively. The continuous random models for amorphous structure, of the type developed by Zachariasen (Zac 32), Bernal, and others

(Be 59, 60a, 60b, Ow 70a), seem to be uniquely different from crystal structures. At the other extreme, there are the models based on the idea that the amorphous solid is an assembly of randomly oriented microcrystallites. For the microcrystallite models to be meaningful, it appears that the crystallite dimension should equal or exceed two unit cell dimensions. At this dimension, most of the material in the system would lie on crystallite boundaries, and the atomic configurations across these boundaries would be more important than those within the crystallites for the overall description of the amorphous structure (War 37). Model studies indicate that the atomic configurations connecting highly misoriented crystallites are quite similar to some of the configurations in the continuous random models; for example, pentagonal arrangements are often seen. This suggests the interesting possibility that the structure of a microcrystallite assembly might degenerate to a continuous random structure when the crystallite size falls below a certain limit.

Dense Random Packing (drp) of Hard Spheres

The coherence of those amorphous solids, with which this study is primarily concerned, is due mostly to covalent bonding, as in amorphous germanium; a mixture of covalent and van der Waals bonding, as in amorphous selenium; or a mixture of covalent and ionic bonding, as in the soda–lime–silicate glasses. However, it may be instructive to consider first the nature of the Bernal dense random packed structure (drp structure) of uniform hard spheres. This structure has a density about 86 percent that of crystalline close packing. It has been characterized by the distribution of its Wigner-Seitz cells (voronoi polyhedra) amongst a small group of ideal forms from which the actual forms of the cells can be derived by small distortions. From this standpoint the structure can be viewed as an admixture of crystallographic cells and noncrystallographic cells (such as pentagonal dodecahedra). The unique feature of the structure is these noncrystallographic elements. When short-range interatomic interactions dominate, as in the condensation of attracting uniform hard spheres, packing to form tetrahedral holes (e.g., rather than octahedral) will be preferred. This should almost always lead to a randomly packed structure, an expectation that was confirmed by Bennett (Ben 70) in a study of computer-generated hard-sphere structures.

STRUCTURE OF COVALENTLY BOUND AMORPHOUS SYSTEMS

In covalently bound systems, the analog of the DRP structure is the random network type of structure that was first proposed by Zachariasen (Zac 32). Recent model studies have shown that these structures can account remarkably well for the pair distribution functions, densities, and configurational entropies of tetrahedrally coordinated amorphous systems. The models are constructed according to the following general procedure (Eh 70): (1) the number of nearest neighbors, their average spacing, and the dispersion of these spacings around the average is made the same as in the corresponding crystal; (2) a certain distribution of distortions of the bond angles from their ideal crystal values is allowed; (3) the surface density of dangling bonds is kept constant during the building of the model. In this way an "ideal" amorphous structure is formed that can be enlarged indefinitely without the development of prohibitive strains. It has been shown (Bel 66, Ev 66) that such a model satisfactorily accounts for both the pair distribution function (PDF) and density (97 percent of crystalline) of fused silica with average distortions of $\pm 15°$ (maximum $\pm 30°$) from the average Si–O–Si bond angle taken to be $150°$. Bell and Dean also showed (Bel 68), by considering the options available in enlarging the model according to the described procedure, that the configurational entropy should fall within the range of the observed values (ca. $\frac{1}{4}$ k/molecule) for the transition from cristobalite to fused silica. The described rules were used by Polk (Pol 71) to build a random network structure for the tetrahedrally coordinated elements (see Figure 1). Average distortions of $\pm 10°$ (maximum $\pm 20°$) above the ideal tetrahedral bond angle were allowed. The structure so formed had a density 97 ± 2 percent of that of the crystal, and its PDF is in excellent agreement with that of amorphous silicon as determined by Moss and Graczyk (Mos 69). The local configurations in random network structures can also be characterized by a small number of ideal forms, in this case rings, from which the actual forms can be derived by small distortions. As with the random sphere packing, it is found that the amorphous structure viewed in this way is an admixture of noncrystallographic and crystallographic elements; for example, the Polk structure for amorphous Ge and Si can be described as an assembly of 5- and 6-membered rings. Similar structures have been generated recently by computer (Sh 71, Hen 71).

Perhaps the most interesting and significant feature of these random network structures is that, in contrast with the DRP hard-sphere structure (density 86 percent that of crystalline close packing), they exhibit

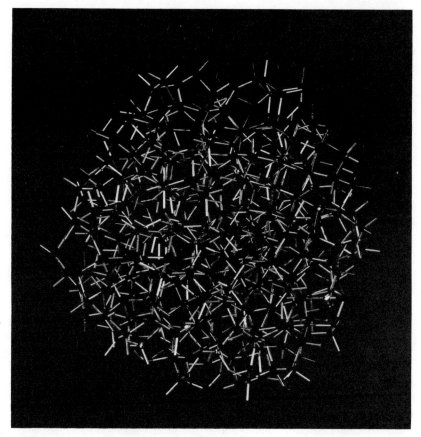

FIGURE 1   Random network model for an ideal amorphous structure of a tetrahedrally coordinated element (Pol 71).

densities closely approaching those of the corresponding crystals. If the energies associated with distorting the bond angles are not too large, this means that their energies will be only a little larger than those of the crystals, as is indeed observed. The question of the dependence of energy on bond angle is an important one for the relative stability of amorphous Ge and Si, and it merits further investigation.

The random network structures that have been discussed, containing no internal dangling bonds, represent ideal structures, which may be more or less closely approached by actual amorphous structures, and which might be the end states of the thermal relaxation processes discussed above. Depending upon their conditions of formation, the actual structures may contain considerable numbers of internal dangling bonds and voids. Even so, the structure of the greater part of the amorphous body might approximate the ideal according to a "swiss cheese" model (Eh 70).

A somewhat different view of the amorphous structure of the tetrahedrally coordinated elements had been proposed by Grigorivici and coworkers (Gr 68, 69b). In this model the 5- and 6-membered rings are little distorted from their ideal forms and incorporated into the structure so that only "staggered" and "eclipsed" configurations of two connecting tetrahedra appear. This scheme would appear to minimize any energy due to bond distortions, but it does not permit indefinite enlargement of the structure without prohibitive strains. The actual structure would have then to be an assembly of more or less discrete amorphous clusters or "amorphons" containing a high concentration of dangling bonds. Also, it appears that there must be a substantial density deficit associated with such an assembly, but this problem seems to have been largely neglected.

The problem of the inter-domain boundary volume contribution is also a troubling one in the application of the microcrystallite models (War 37). It was considered by Cargill (Car 70) when he tested these models with the structures of metallic glasses, but it appears to have been largely ignored in the interpretation of covalently bound glass structures in terms of the model. Cargill showed that a deficit of $\frac{1}{4}$ to $\frac{1}{3}$ monolayer at the boundaries would lead to a very large decrease in bulk density at the crystallite sizes required to account for the x-ray interference function.

Experiment suggests that in amorphous semiconductors generally, as well as in tetrahedrally coordinated systems, the nearest neighbor coordination required by the generally accepted chemical valence, usually as specified by the 8-N rule, is mostly realized. The "ideal" covalent amorphous structure is considered to be one in which *every* atom is bonded to the proper number of nearest neighbors to satisfy its valence requirements. The definition may also include limits on the permissible deviations of bond lengths and bond angles from their crystalline values. An actual amorphous material will, in general, contain, besides the defects noted earlier, a number of unsatisfied valences.

The group V elements, e.g., As and Sb, can crystallize into a 3-coordinated network in which each atom is at the apex of a pyramid formed by the bonds to the three atoms with which it coordinates. These groups are bound together in puckered 2-dimensional layers that are stacked, partly by van der Waals binding, to form the crystal. A random network, which can be a fully connected 3-dimensional one, can be formed from such a 3-coordinated system by distortions of the ideal

A–A–A bond angle (about 98° for the arsenic structure). The As_2S_3 and As_2Se_3 amorphous structures might be regarded as having been generated by the insertion of S or Se atoms between each As–As closest pair of the amorphous As structure. The SiO_2 and Si amorphous structures may be related similarly.

In 2-coordinated systems, such as Se and Te, an amorphous structure can be formed without any bond distortions by rotation of neighboring chain segments randomly relative to each other. In this way, a long chain takes the form of a random coil. In an actual system, these coils will interpenetrate and there will be strong van der Waals interactions between neighboring segments of a coil. It is now believed (Lu 69) that in addition to coils, amorphous Se contains quite a large admixture of rings, primarily 8-membered, which are bound into the structure by van der Waals forces, and possibly by interlocking with coils.

Mott (Mo 67b) suggested that in amorphous semiconductor solutions, in contrast with crystalline solutions, the chemical valence of each constituent atom is everywhere satisfied. This view seems to be supported by most of the experimental evidence. In glasses formed by slow cooling of melts, there probably is enough time for the achievement of this chemical valence satisfaction when it is energetically preferred. However, as Mott noted, this ideal chemical bonding might not be fully realized under some conditions of amorphous solid formation by vapor quenching. Strong evidence for the attainment of local valence satisfaction by covalent bonding was obtained for the Te–Ge amorphous system by Bienenstock and coworkers (Bi 70). In the Te-rich alloys, the tetrahedral coordination of the Ge results in the cross-linking of the Te chains into an amorphous 3-dimensional network.

Phase Separation and Crystallization of Amorphous Systems

To this point we have considered only the formation and structure of a single amorphous solid phase. As we noted, such a phase is not the thermodynamically most stable one, and so it may evolve in a variety of ways into a thermodynamically more stable polyphase system. A complete characterization of the structure then requires information about the spatial distribution of the several phases, as well as of the molecular configuration within each phase. To discuss this characterization, it will be helpful to define the various temperatures that are pertinent in the structural evolution. The glass temperature, T_g, has already been defined. The thermodynamic crystallization temperature, T_{tc} (often designated T_m), is the temperature at which the liquid co-exists in equilibrium with one or more crystalline phases; it lies well above T_g. Liquids containing two or more components often are prone to separate into two liquid phases. The temperature at which phase separation becomes thermodynamically possible at a given composition will be denoted by T_{tp}. Often liquid immiscibility gaps open only in the temperature range where the liquid would be undercooled; i.e., T_{tp} falls well below T_{tc} for many liquids.

The isothermal time constants for both crystal growth and phase separation in covalently bound systems are usually found to scale roughly as the shear viscosity (Tu 69a). Consequently, the rate of structural evolution in an amorphous system usually becomes very sluggish and often imperceptible at temperatures below T_g. Crystallization of amorphous phases may, under certain conditions, occur fairly rapidly at temperatures between T_g and T_{tc}. Sometimes a term, kinetic crystallization temperature, T_{kc}, is used to denote the temperature at which the crystallization rate becomes very rapid. However, this term can be quite misleading, since the crystallization rate depends critically on the seed (nucleation center) density and may be substantial over a considerable temperature range (Tu 69a). It is probable that most covalent amorphous solids, especially films on substrates, already contain a considerable density of nucleation centers so that their crystallization may be governed primarily by the crystal growth rate, u. This rate is generally described as the product of two factors, $u = f_1 (\Delta T)$ $f_2 (T)$; one, $f_1 (\Delta T)$, is a thermodynamic factor that increases at a moderate rate with the undercooling ($T_{tc} - T = \Delta T$), and $f_2 (T)$ is a kinetic factor that decreases sharply with decreasing temperature, T (Tu 69a). In covalently bound systems, the possibility exists that the kinetic factor in crystal growth can be sharply increased by extraneous effects such as trace impurities (Tu 58) or photon absorption, which might lead to the breaking of covalent bonds. Such effects seem fairly well documented in the crystallization of amorphous selenium. For example, halogen additions markedly reduce the viscosity of liquid selenium and increase the crystal growth rate (Ke 67), presumably by breaking bonds and thus reducing the length of the selenium chains. Also, Dresner and Stringfellow (Dr 68) have demonstrated a marked photoenhancement of the crystal growth rate in amorphous selenium. More recently, Feinleib, et al. (Fe 71), have reported a marked photoenhancement of the crystallization rate of Te–Ge-based glasses.

As we have noted, a liquid consisting of two or more chemical components often has a very strong thermodynamic tendency to separate into two liquid phases when it is undercooled to the vicinity of its glass temperature. The reason for this is that T_g is quite low, relative to the coherence energy of the system, so that there is little entropic stabilization of the glass solution. Further, most glass compositions will not correspond to any of those most favored energetically (Morr-). These are likely to be the pure components or a pair of ordered solutions, each with some simple stoichiometric ratio (e.g., As_2Se_3 or $GeTe_2$). From this point of view, as was shown by Morral and Cahn (Morr-), the thermodynamic tendency toward some kind of phase separation will be greater the more components there are in the solution. Further, the interfacial tension between two amorphous phases is relatively small and it vanishes altogether at the consolute (critical solution) temperature (Ca 68). Consequently, at wide departures from equilibrium, phase separation in amorphous solutions can occur on a very fine spatial scale; e.g., a few tens of angstroms. Since the distances over which diffusion must occur are so very small, phase separation is very rapid and difficult to suppress even when the viscosity of the system is as high as, for example, 10^6 to 10^7 poise. However, at this viscosity the phase-separated structure, once formed, is likely to persist (i.e., gravity segregation of the separated phases will be very slow) when the system is cooled into the glass state. This means that there is a high probability that any multicomponent glass formed by cooling its melt will be separated into two phases, often interdispersed on a very fine spatial scale. *Indeed, this phase separability characteristic is, perhaps, one of the most unique and valuable properties of multicomponent glass-forming systems.* It can be and has been exploited to achieve phase interdispersion with much smaller periods than is possible by other methods. Further, it often happens that one of the liquid phases separating in this process can crystallize very rapidly because it is more fluid and/or more undercooled than the parent liquid (Mau 64). This leads to a body in which one of the interdispersed phases is crystalline. When phase separation occurs by a spinodal mechanism (i.e., by amplification of periodic composition fluctuations) (Ca 68, Hil 68), the two phases will initially be interconnected (Ca 65). Phase interconnectivity also can be established, following a nucleation and growth-phase separation, by a coalescence process (Hal 65, Se 68). The sequence of processes considered here may be summarized as follows:

$$\text{Homogeneous viscous liquid} \xrightarrow[\substack{\text{phase}\\\text{separation}}]{T < T_{tp}} \left.\begin{array}{c}\text{A liquid}\\+\\\text{B liquid}\end{array}\right\rbrace \longrightarrow \begin{array}{c}\text{A glass}\\+\\\text{B crystal}\end{array}$$

The rate of phase separation increases quite rapidly as $T - T_{tp}$ increases. If one of the phases is easily crystallizable, it may appear that the initial kinetic crystallization temperature is almost indistinguishable from that at which phase separation occurs.

III

Preparation

The basic goal in preparing amorphous materials is to freeze them into a metastable state characterized by the absence of long-range order. This can be accomplished by a large number of methods that can be grouped into two basic categories. One is to introduce disorder by thermal methods, then to quench from the liquid or vapor to below T_g sufficiently rapidly to prevent the achievement of internal configurational equilibrium. The other is to create the disorder in a solid below the temperature at which it can regain long-range order in the time scale of the experiment. Different methods may or may not give equivalent films for a variety of reasons that are often difficult to determine.

While the causes of nonreproducibility among samples may be determined by application of the characterization techniques discussed in the next section, it is during the preparation steps that it is created and the possibility of control exists. Although glasses may approach an "ideal" metastable noncrystalline state (Section II), they can be quenched into many other metastable states, making structure and properties preparation dependent. Control over composition and purity is also poorer than with crystalline materials since the potential for impurity rejection or leveling during crystal growth is lost.

The glasses of interest here represent a broad range of preparative challenges. Se and the As–Se glasses are easy to form by quenching

16

and have been prepared from the melt and vapor. Experimental evidence indicates that the quenched glasses retain the local coordination of the liquid, which, in turn, is closely related to the molecular structures of the crystalline forms (My 67). Ge has not been quenched from the melt. X-ray studies (Kr 69, Br 71b) show the glass structure to be closely related to the crystalline form in contrast to the liquid which has very low viscosity and shows metallic conduction. Amorphous Ge has been prepared by (Br 71b) evaporation, sputtering, glow discharge, electrolysis, ion implantation, transformation of a high-pressure-crystalline polymorph (Bu 71), and phase separation from a Ge–GeO$_2$ solution (deN 71). While a-Ge films prepared by various techniques appear to be structurally similar, their properties vary widely.

QUENCHING

The most common preparative method is quenching from the melt or vapor. Melt quenching rates extend from $\sim 10^{-2}$ °C/second in an annealing furnace, to 10^3–10^4 °C/second in strip furnaces, to 10^5–10^7 °C/second by the more complex splat cooling techniques (Sa 68). Vapor quenching rates overlap the high end of this range, and rates as high as 10^{15} °C/second have been reported (No 69). The choice of method is usually dictated by the material of interest, since the faster quench rates are achieved at greater experimental complexity and cost. The number of nucleii and crystal growth rate determine the minimum required quench rate (Tu 69a).

Vapor Deposition

Vapor deposition is the most commonly used technique for materials considered in this report. A number of special techniques have been developed and will be mentioned here in the context of advantages or disadvantages in the preparation of amorphous semiconductors. Numerous reviews on thin-film preparation are available for further information on specific systems, techniques, and materials (Ch 69) (Mai 70).

All vacuum deposition systems consist of several basic elements, a vacuum chamber, a source of the material to be deposited, a substrate and associated fixturing. A most significant factor is the amount and nature of contaminants, including the ever-present atmospheric gases, available for incorporation into the material of interest. All pumping systems except cryosorption contribute some foreign material, e.g., hydrocarbons, Ti, and Hg. Fortunately, mercury pumps are rarely if ever used in this application since some materials, such as Se, are excellent getters for Hg vapor. In addition, systems vary in their pumping speed for various atmospheric gases, in effect concentrating certain species. The surface of the film being deposited is exposed to sufficient background gases to condense ~ 4 monolayers per second at 10^{-5} Torr, and at 10^{-9} Torr $\sim 4 \times 10^{-4}$ monolayers/sec. What, if any, material is incorporated, how it is incorporated, and its effect must be individually considered. It is frequently noted that it becomes more difficult to quench an amorphous film at higher vacuums, indicating a stabilizing effect of impurity incorporation.

The substrate temperature during sample preparation is a particularly important parameter. Too low a temperature results in low-density films with poor adhesion. At higher temperatures, there may be sufficient mobility only to allow complete replication of the substrate, while slightly above this, there is sufficient flow to provide very smooth surfaces and higher densities. At still higher temperatures, crystallization begins and the ability to quench an amorphous phase is lost. Since physical and molecular structure can be affected by quench rate and annealing, other measured properties may vary with substrate temperature at preparation.

The classical method of providing a vapor of the desired material at the substrate is evaporation. The sophistication comes in the choice of heating methods. Simple resistance or RF induction heating of a source is appealing because of its simplicity, but introduces problems of contamination and fractional distillation (Ef 69) in multicomponent materials. The first problem can be minimized by proper choice of crucible material or eliminated by using the material as its own crucible. The most common method for this is electron bombardment, where a focused electron beam causes evaporation from a small heated region on the surface of a larger piece of the material of interest. Such localized heating can also be accomplished by energy absorbed from a laser source that is physically located outside the vacuum chamber. If the source material is sufficiently electrically conducting, exploding wire and arc methods are also possible although more commonly used for metals and refractory materials. These methods all subject the evaporating material to high energies and high local temperatures, which may produce different vapor species than simple thermal sources.

The problem of fractional distillation is commonly circumvented by coevaporation or flash evaporation techniques. In coevaporation, different components are fed into the vapor stream from separate sources. Stoichiometry is controlled by the temperature or surface area of each source. Feedback control of deposition rate by source temperature is possible, but independent source calibrations may not be valid due to vapor phase reactions and different accommodation coefficients of the actual film. While uniform distribution of the major components can be achieved, impurities may still fractionally distill from each source resulting in their concentration at one or both film surfaces.

In flash evaporation, material is continuously fed into a source at a rate slow enough to prevent the buildup of a pool of molten material so that the instantaneous average composition of the vapor is that of the feed material. This method is slow, inefficient, and usually results in films with numerous defects due to spatter of solid and liquid material from the source. A variation uses continuous feed to a pool of molten material with the rate controlled so that the vapor is of constant composition, although different from that of the feed material.

The source temperature controls the evaporation rate and may affect the vapor species and, by means of radiant heating, the film temperature. The condensation rate is also dependent on the evaporation rate but subject to some independent control through substrate temperature. Film properties may be affected by the deposition rate in several ways. Fast deposition favors low-density films since there may be insufficient time for surface mobility processes, while slow deposition allows more time for reaction with and incorporation of residual gas species present in the vacuum chamber.

Sputtering is another method that can avoid fractionation effects and is finding ever wider application, including commercial use at ECD (Nea 70b). The surface of the material to be deposited is the target of bombardment by energetic ions which sputter target material free to be collected on the substrate. The ions can be inert, Ar being the most popular, or reactive where the desired deposit is a compound of the target material, and a second component such as O or S is introduced with the sputtering gas.

The basic characteristics that make sputtering attractive are many. A clean substrate can be prepared by using it as a target and sputtering away surface contaminants before deposition. The sputtered species are themselves highly energetic and provide dense and highly adherent films, providing the target is maintained at a sufficiently low temperature to prevent sublimation processes. Uniform film composition matching that of a multicomponent material target is the rule rather than the exception, although target bulk diffusion, decomposition, surface reaction, or widely varying sticking coefficients on the substrate, can cause problems. Large areas of uniform thickness can be prepared with high efficiency, although there may be problems with target preparation. Targets should have an area about twice that of the film desired. The deposition rate is easily kept constant and can be controlled by gas pressures and accelerating voltage.

The method is, of course, not without disadvantages. The sputtering gas provides an additional potential film contaminant as well as a contaminant carrier. Although avoiding the localized intense heating of a thermal source, the plasma discharge heats large areas with resultant outgassing. The high-energy species lead to new contaminants through reaction with other parts of the system and cracking of hydrocarbons. Deposition rates are slow compared to thermal methods. The additional cost and complexity of the system may also be a factor.

Many variants of the basic process have been developed. The substrate can be DC- or asymmetrically AC-biased to provide some bombardment cleaning of absorbed gases, which would otherwise be trapped at an obvious loss in efficiency. The systems normally operate at 20–100m Torr to sustain a discharge. Cleaner films can be obtained by sputtering at lower pressures using a magnetic field to increase ionization efficiency, an auxiliary source of electrons or ions, radio frequency excitation, or a combination of these. Reasonable deposition rates at pressures as low as 10^{-4} Torr are possible. RF sputtering also provides the cleanest method of removing the surface charge from an insulating target making sputtering of insulators possible and significantly increasing the rate for low conductivity materials. Even under optimum conditions, sputtering is a relatively slow process. Rates are generally less than 10 Å/sec compared to 10–1,000 Å/sec for thermal sources.

Chemical vapor methods requiring heat, such as vapor phase pyrolysis or highly exothermic reactions, may not be suitable because unavoidable heating of the substrate and depositing film results in crystalline deposits. Thermal pyrolysis of silane above 800°C is used to grow epitaxial crystalline films of Si. Vapor phase decomposition of silane in a glow discharge occurs at low temperatures and, although slow, results in a-Si films reported to have dramatically different electrical properties from those prepared by other methods (Br 71b).

Liquid Quenching

For ready glass formers, the techniques of melt quenching are relatively straightforward and well documented in the oxide glass literature. Large samples of As–Se, Ge–As–Se, Ge–Sb–Se (Ta 71), and other glasses of excellent optical quality are commercially available. Important precautions include assuring complete reaction and chemical homogeneity of multicomponent glasses, vacuum outgassing, and preventing high-temperature reaction with crucibles, ampul materials, and ambient atmospheres. Sample mass and geometry, heat capacity, and thermal conductivity control the achievable quench rate. The basic parameters involved and methods of predicting required quench rates have been studied but will not be further discussed here (Sa 68, Tu 69d). However, when reporting quenchability of a glass, actual cooling rates or data defining sample mass, surface area, etc., should be reported since the ease of quenching and, therefore, definition of glass-forming compositions is sample dependent.

Splat cooling (Du 70) is less frequently used since it often produces samples that are strained and of an unsuitable geometry and perfection for most measurements.

OTHER METHODS

Solids can be transformed to a disordered state in a solid-state reaction with the energy provided by radiation (neutron, α particles, etc.), shear, or chemical reaction in processes often referred to as amorphization (Ro 70). The chemical reaction need not be completely solid state; in fact, reaction with, or evolution of, a vapor in an oxidation, reduction, or disproportionation reaction is often involved.

A variety of other chemical methods is also used to prepare glasses. In addition to the vapor and vapor–solid reactions previously mentioned, there are numerous solution processes such as electroplating, electroless plating, anodization, and polymerization. In general, films prepared by chemical processes are subject to wide property variations due to trapped impurities, poor stoichiometry control, and inhomogeneities resulting from incomplete reactions.

SAMPLE ENVIRONMENT

In addition to the preparation of the amorphous layer, one must consider the total sample, including substrate, electrodes, and free surfaces.

The choice of a substrate is generally determined by the measurement to be made but must be consistent with a set of secondary restraints. Matching thermal expansion coefficents is important, especially if measurements are to be made as a function of temperature or if deposition is at a temperature far from ambient. Glasses are brittle below T_g and will crack easily, especially under tension. Sample flow may also occur during measurements above T_g, resulting in lack of reproducibility due only to a new film thickness.

A second consideration is chemical reactivity. Reaction of the depositing vapor or liquid with the substrate or adsorbed films, or slower diffusional processes, can result in alloy or compound formation. In addition, gas diffusion through the sample and reaction with the electrode is possible. If electrical measurements are to be made, the electrodes, including the substrate if it serves as an electrode, may be chosen for their blocking or injecting characteristics. Chemical reactions with the layer or diffusing gases may result in an electrode with entirely unexpected behavior (Ut 71). It has recently been suggested (Br 71a) that only noneutectic forming metals should be used for electrodes and that much earlier data need reexamination. The substrate can also serve as a source of nucleation sites for crystal growth in the sample.

Chemical reaction of a free surface with the ambient atmosphere is another possibility. The surfaces of As–Se films have been shown (Tr 69) to oxidize in air at room temperature, yielding crystalline As_2O_3 and a surface glass layer rich in Se.

One must never forget that these materials are in a metastable state with respect to a crystalline or phase separated form. Transformations may occur slowly but continuously without external stimulation, or more rapidly when external influences are present. The crystallization of Se in the presence of light (Dr 68) or water vapor (Chi 67) is one example. Many problems can be avoided by storing samples at low temperatures, in the dark and in an inert atmosphere. Encapsulation may be considered where practical.

IV

Characterization

The importance of materials characterization has been recently addressed in the report of the MAB Committee on Characterization of Materials (MAB 67). It states,

Clearly, much of solid-state research is concerned with the effort to understand properties, in terms of a particular composition and structure, *but it cannot be emphasized too strongly that a substantial fraction of this effort is of marginal value because it is carried on without a clear understanding of the true nature of, and need for, characterization, and it fails to distinguish between property studies on characterized and uncharacterized materials.*

One has only to leaf through a collection of current papers in the field of amorphous semiconductors to be convinced that the problem exists here. While a majority of the experimental papers gives some preparative information, few report any attempt at characterization outside the specific measurement to which the paper is addressed, and certainly none to the standards proposed in the MAB report. It is characteristic of the field that in approximately 2,000 pages (Ch 69, Mai 70) reviewing the preparation and properties of thin film, less than 40 are devoted to structure and composition, and nearly half of these discuss diffraction techniques not generally applicable to amorphous films.

This lack of attention is not surprising for several reasons. Materials characterization is difficult, time-consuming, and expensive. Existing characterization tools and sampling techniques may not be directly applicable and, in some cases, new tools may have to be developed. Much work in this field is carried out presently by individual investigators in academic laboratories, while an industrial scientist in a more established field may have the direct support of as many as 2 or 3 other workers in materials preparation and characterization.

The MAB Committee evolved as a working definition (MAB 67):

Characterization describes those features of the composition and structure (including defects) of a material that are significant for a particular preparation, study of properties, or use, and suffice for the reproduction of the material.

A particularly important fact is that major attention to amorphous semiconductors is recent and consequently it is usually unknown what compositional and structural features directly and unambiguously define a material. While it would be of great value for this report to specify a minimum characterization to insure reproducibility, this is not yet possible except in a limited way for a few specific materials. Certainly, any characterization must address the three important questions of chemical composition, whether or not the specimen is truly and completely amorphous, and whether or not there are multiple phases present. In the latter case their composition and distribution must be specified. In addition, while glasses may approach an "ideal" metastable noncrystalline state, they can be quenched into many other metastable states which may be changed by annealing. Measurement of the true bulk density or the heat capacity over a very limited temperature range can serve to characterize a sample without altering it. Measurement of thermal characteristics over a wide temperature range by, for example, differential thermal analysis (DTA) or scanning calorimetry (Mu 70) can yield significant further information but may be a destructive test causing crystallization or phase separation or allowing the sample to relax to a new state. These are relatively easy, but important, measurements and provide necessary data to test structure models.

The practical question of what must be specified to insure reproducibility deserves comparable attention to the more fundamental determinations of composition and structure. An example of evolving property–structure–composition relationships is provided by the current controversy over the properties of a-Ge and Si. A large part of the early nonreproducibility is now attributed to differences in deposition temperature or post deposition annealing. A model has evolved (Br 71b), largely from radial distribution functions, small angle x-ray scattering, density and ESR data, which incorporates small voids into an otherwise dense amorphous matrix. The effect of voids on electrical and optical properties is modified by annealing. It is not yet clear just how annealing influences the voids since the gross film density does not change (Th 71). While further structural studies are conceived and contemplated to understand these effects, other experiments can be carried out and reproduced by carefully controlling and reporting deposition conditions and subsequent annealing.

The remainder of this section will concentrate on composition and structure, tools available for their measurement, and some specific problems associated with the amorphous semiconducting materials.

COMPOSITION

Although questions concerning molecular species, valence states, and clustering might be considered compositional, they will be discussed in the part devoted to structure below. The principal concern here is chemical purity and stoichiometry.

In the literature, the question of sample purity is often ignored, or dismissed with a statement concerning the use of 5-9's or 6-9's pure elements. This may in part be due to the widespread belief that amorphous materials are impurity insensitive. In some instances there is good evidence to support this belief. There are also many reports of the dramatic effects of a few parts per million (ppm) to a few percent of impurities, for example the electrical effects of S, Te, As, Cl, Tl, Bi, Ge (Sc 70), and O (La 70) in Se, the optical effects of O in Se (La 70), As–Se (Va 70), As_2Se_3 and GeAsTe (Sav 65), SiAsTe (Hi 66), Ge (Ta 71) and GeSe (Va 70), the physical effects (e.g., T_g, microhardness) of I and Ge on As_2Se_3 (Ko 62), and modification of the crystallization rate of Se by Sb and Te (Dz 67), to list a representative few.

Use of the designation 6-9's purity may also be misleading since for most elements it is of questionable validity. To establish that a material contains less than 1 ppm total impurities requires analysis for about 100 elements to a sensitivity and accuracy of less than 10 ppb each. The technique most commonly used is emission spectroscopy, which can detect approximately 70 elements with a sensitivity of the order of 1–10 ppm. Even if materials of this purity are obtainable, maintaining this level through sample fabrication is also demanding and not generally addressed unless proven necessary.

Elements common to the amorphous semiconductors considered in this report range from Ge and Si, which can be purified to these high levels, to Se, which is relatively impure. Selenium purification processes are limited by an inability to zone refine, ease of compound formation including organoseleniums, and the ease of generation and extreme toxicity of H_2Se. In addition, fractional distillation during thin film preparation and slow diffusion during melt quenching may result in local concentrations of these impurities, especially at surfaces, where their concentration is sufficient to affect property measurements but too low on average to be detected by conventional chemical analyses.

The importance of chemical analysis cannot be overstated. Experimenters interested in preparation or property measurement should be aware of the actual purity range of the starting materials, the importance of impurity effects in related materials or measurements and the available analytical tools which might give meaningful information. Little can be said here about the actual methods. Chemical analysis is an active field with methods ranging from older established wet chemistry through ever more rapid and reliable instrumental techniques. Reviews of methods, their precision and sensitivity, and unique features are available. For example, the state of the art is reviewed biennially (even years) in a special issue of Analytical Chemistry (AC 70). In general, the application to amorphous semiconductors is straightforward. If the glass is easily melt quenched, large samples can be available for analysis. In the case of thin films or splat-quenched glasses, only a few milligrams may be available, and simple compositional analysis takes on the complexity of a trace element determination in macro samples. Such samples also have very high surface areas, and bulk analyses may be difficult because of the large influence of surface contamination.

The determination of compositional gradients on a micron or submicron scale, for example, those resulting from distillation effects in thin-film preparation by vacuum evaporation (Ef 69), or local inhomogeneities due to phase separation, also challenge the state of the art. Recent developments such as the electron probe microanalyzer (microprobe) (Cam 70), ion microprobe massanalyzer (IMMA) (Evan 70), microfocus x-ray (milliprobe) (Cam 70), and Auger spectroscopy (Harr 68) are having a major impact and hold considerable promise for the future.

STRUCTURE

Structure, as defined here, ranges in scale from macroscopic defects (e.g., cracks and bubbles), which are important if experimental data are to be meaningful, to atomic bonding and molecular structure, which are more important for theoretical understanding and modeling. Phenomena such as crystallization and phase separation can span the range from visible to the unaided eye to the Angstroms limit of the most powerful microscopes available. Obviously no single tool can be recommended for all characterizations.

Macroscopic *(≥ 10 Å)*

Optical microscopy (Coc 70) remains a primary tool. Gross defects such as bubbles, cracks, foreign inclusions, surface contamination, and reaction products, as well as crystallization and phase separation down to about 1 μ, can often be easily identified. For other than very thin films or surface effects, an infrared microscope is required for many materials (Va 68). The scanning electron microscope (Fish 70) overlaps optical microscopy and extends its range down to less than 100 Å. Using the standard secondary electron mode, surface defects can be studied throughout this range. Other modes of operation, such as conduction, backscattering, and cathodoluminescence, can give additional information on surface and/or bulk properties (Fe 71). Density is another measurement sensitive to the presence of voids or inclusions, and is capable of quite high sensitivity. It cannot alone, however, describe the nature or size of the inhomogeneity.

Phase separation may involve a crystalline phase of the same or different composition or a second amorphous phase of different composition. The former is usually determined by x-ray (Pf 70) or transmission electron diffraction (Fish 70). Reflection electron diffraction can reveal the presence of a crystalline phase on the surface typical of the bulk or possibly due to a reaction product such as an oxide. Selected area diffraction, taking advantage of a focused x-ray or electron beam, may offer further sensitivity advantages. These techniques not only establish the presence of a crystalline phase but can generally give positive identification as well. As the size of the individual crystallites decreases and/or strain increases, the diffraction peaks spread. In the 20–50 Å region, interpretation is controversial.

The presence of two or more amorphous phases may be detected by optical or electron microscopy. The advent of high-energy electron microscopes (MeV) makes transmission electron microscopy applicable to thicker films without requiring sample thinning procedures that may themselves affect the structure. However, identification of the phases is generally accomplished through high resolution x-ray fluorescence. Current electron probe microanalyzers (Cam 70) can do quantitative chemical analysis, for all elements with atomic number 5 or greater, on a micron scale. The scanning electron microscope can also be equipped for x-ray analysis and offers somewhat higher resolution. Auger spectroscopy (Harr 68) offers high sensitivity for the light elements. Small angle x-ray scattering is a technique, frequently applied to polymers, which can detect repeat distances of 10–10,000 Å and thus can characterize phase separations with domains in this size region. If the two phases differ significantly in other properties such as conductivity, optical absorption or reflectivity, dielectric constant or loss, etc., then detection and identification may be accomplished more readily or with greater sensitivity by methods based on such a property.

Microscopic (≤ 10 Å)

The greatest handicap to structural characterization of amorphous materials is that the absence of long-range order precludes complete structure determination by standard x-ray, electron, and neutron diffraction methods. The scattering which is observed can be used to determine a radial distribution function (RDF) from which the number and distance of at least the first and second nearest neighbors of an average atom can be determined. Recently Mozzi and Warren succeeded in greatly improving the resolution of the x-ray method by using procedures that minimized the Compton modified scattering (Moz 69, 70). Even if the RDF calculated for a model structure matches the experimental RDF, there is still no guarantee that the model structure is unique. At least a dozen studies of a-Se can be found in the literature. Using similar experimental data, nearly half that number of different structures including 2- and 3-dimensional chains, 8–500 atom rings, 6 atom packets, and combinations of these have been proposed (Gr 69a, Bre 69). Similarly, proposals have been made to the effect that the structural units of a-Si involve distorted tetrahedra, pentagonal dodecahedra, or diamond-like microcrystallites (Mos 69). a-Se is found to have 2 neighbors at 2.34 Å and 8 at 3.72 Å, identical to both trigonal and monoclinic Se; and a-Si 4 at 2.35 and 12 at 3.86, again identical to the crystal form. Therefore, one must depend on data at larger interatomic distances where resolution is poor due to the overlap of many peaks and is also subject to large experimental errors. In the case of a glass with two or more elements of significantly different scattering power, for example, GeTe, it is possible to gain additional information with respect to a preference for Ge–Ge, Ge–Te, or Te–Te bonding (Bi 70). In general, however, it is safer at present to complement RDF data with information from other experimental techniques to choose among possible structures.

Additional information about nearest neighbors can be obtained by any effect characteristic of the chemical bond involved. Principal among these are the various forms of optical spectroscopy (Cr 70). There is often sufficient carry-over of spectral features from the crystalline form or forms to permit the formulation of convincing arguments. IR has been used (Lu 69) to determine the presence of both ring and chain molecules in a-Se corresponding to the monoclinic and trigonal crystal forms respectively. Raman spectra of As–S glasses (In 69) have shown the presence of S–S and As–S bonds in both the molecular forms found in Realgar [As_4S_4] and Orpiment [As_2S_3]. Electron (He 70) and x-ray spectroscopy (Cam 70) are capable of describing the atomic bonding of the elements present and can also be useful in these determinations.

Localized unpaired electrons can be detected by, for example, ESR (Kon 70) or static susceptibility (Ta 70b) measurements (Mul 70) and may provide useful information. NMR (Hee 70) can also describe the environment of atoms of those elements for which probes are available (e.g., ^{75}As, $^{121,123}Sb$, ^{209}Bi, ^{35}Cl, ^{77}Se) (Pe 68).

The question of molecular species, particularly macromolecules, has been addressed by numerous indirect measurements. While these are not usually considered as part of the characterization procedure, property measurements have been shown to be directly related. The presence of Se_8 rings and Se_n chains can be demonstrated using IR spectra. Their relative abundance and the average chain molecular size have been determined by differential solubility (Bri 29), viscosity vs. halogen concentration (Ke 67), magnetic susceptibility (Ma 64) and Raman spectroscopy (Wa 69). In some binary systems, structural arguments can be developed from thermodynamic (My 67), viscosity (Ne 63), elastic constants (De 69), microhardness (Ko 62), solubility (Bo 69), magnetic susceptibility (Ci 70), and other measurements (Ts 71), made as a function of composition.

The specific use of a material may also suggest further characterization tools such as Schlieren techniques (Va 68) to detect local compositional variation in optical windows and electron beam current measurements to detect local electrical inhomogeneities in vidicon applications.

At the present, it is likely that optimum application of these techniques will still result in a less certain characterization than for comparable crystalline materials (Ka 70). This emphasizes the initial admonition quoted from the MAB report, which was largely addressed to crystalline substances. More rapid progress in understanding the properties of amorphous semiconductors is possible if these characterization methods are intelligently applied, new tools are constantly sought, and the importance of the relation and sensitivity of properties to structure and composition is widely acknowledged.

V

Fundamental Properties of Amorphous Semiconductors

GENERAL CONCEPTS

Because the understanding of amorphous semiconductors is strongly influenced by the body of knowledge concerning crystalline systems, it is appropriate to review first some of the key features of the electronic states in crystalline semiconductors (Kit 66, Zi 64).

1. A perfect crystalline semiconductor at $0°K$ has an empty conduction band of states, separated by an energy gap from a filled valence band. The states in the two bands may be regarded roughly as derived from antibonding and bonding orbitals respectively, but the states are much more free-electron-like than is implied by a tight-binding description.

2. In a pure, nonvibrating crystal, and in the one-electron approximation, the wave functions in the two bands may be written in the Bloch form, i.e., as the product of a plane wave of definite wavevector, $\vec{k}$, and a function having the periodicity of the lattice.

3. In any real crystal there will be defects (e.g., impurities, vacancies, interstitials, dislocations), which scatter the Bloch waves, so that the wavevector, $\vec{k}$, is only an approximate quantum number characterizing the states. Scattering is also produced by phonons, and at high energies

or finite temperatures, by electron–electron interactions. In addition, defects can give rise to states within the energy gap of the perfect crystal.

4. At low temperatures, the Fermi level of a perfect semiconductor lies halfway between the bottom of the conduction band and the top of the valence band. The number of charged carriers (electrons in the conduction band and holes in the valence band), and hence the electrical conductivity at low temperatures, is proportional to $e^{-E_a/kT}$, where the activation energy, E_a, is one half the energy gap, E_g. This form for the conductivity is called the "intrinsic" conductivity of the material. In any real material, however, at sufficiently low temperatures, the very small intrinsic conductivity will be dominated by the contribution of carriers associated with impurities or other defects, which give rise to states in the gap, and generally cause the Fermi level to be closer to one or the other side of the energy gap. The conductivity then depends strongly on the number and kind of defects present, and is called "extrinsic." In general, the more imperfect the material, the larger the value of the extrinsic conductivity, and the larger the temperature range over which the extrinsic conductivity is seen.

5. The principal optical absorption band in a crystalline semiconductor is associated with transitions of an electron from a state in the valence band to a state in the conduction band having the same wavevector as the initial state. The fundamental optical absorption edge occurs at the gap energy, E_g. For crystals where the lowest state in the conduction band occurs at a different point in the Brillouin zone than the top of the valence band, optical absorption near the fundamental edge requires the aid of a phonon or crystal defect to conserve wavevector. In a real crystal, absorption well below the threshold, E_g, can occur because of transitions from a filled defect level inside the energy gap to the bottom of the conduction band, from the top of the valence band to an empty defect level, or from one defect level to another. In a crystal with a small density of defects, one would see a number of sharp thresholds in the absorption, corresponding to transitions between the isolated defect levels and the edge of the conduction or valence band. In a crystal with a large density of defects, or many kinds of defects, one would find only a relatively featureless tail in the optical absorption below the fundamental edge. An absorption tail below the fundamental edge can also result from interaction of the electronic states with lattice distortions (phonons). The position of the fundamental absorption edge is also lowered somewhat by the electron-electron interaction, which leads to a series of electron-hole bound states (excitons) just below the continuum of states beginning at E_g.

Many features of the crystalline semiconductor persist in the amorphous. There will again be a filled valence band, roughly derived from bonding orbitals, and an empty conduction band derived from anti-bonding orbitals. In the amorphous material, $\vec{k}$ is not a good quantum number for the electronic states. Some remnants of $\vec{k}$-conservation may persist, however, in that states in a given energy range may contain predominantly wavevectors associated with particular portions of the Brillouin zone of the ordered structure.

Remnants of $\vec{k}$-conservation in the optical absorption spectrum would be manifest as enhanced matrix elements for optical transitions between states associated with the same portion of the Brillouin zone. Thus, in a crystal with impurities or phonon scattering, transitions that conserve wavevector are generally much stronger than *indirect transitions,* where the initial and final states are associated with different parts of the Brillouin zone. This should be distinguished from *nondirect transitions,* which occur when the remnants of $\vec{k}$-conservation are negligible. In the latter case, the optical absorption reflects a simple convolution of the densities of states of the valence and conduction bands. Nondirect transitions have, in fact, been evoked to explain photo-emission results in a number of crystalline materials as well as in amorphous Ge (Sp 67, Sp 70).

The rough features of the density of states will generally be the same in the amorphous material as in the crystalline, provided the short-range ordering of the atoms is similar for the two cases (Io 60). Sharp features of the crystalline density of states, arising from critical points in the Brillouin zone, where there is a vanishing of the gradient of the energy value with respect to wavevector, would, of course, be considerably smoothed out in the amorphous system. Corresponding to the energy gap in the crystal, there will be a quasigap in the amorphous material, where the density of states is much smaller than in the valence and conduction bands (see Figure 2). Although an ideal covalent amorphous structure might still possess a true energy gap with zero density of states, there will always be a finite state-density in real materials. The states in the quasigap are believed to be localized in space, and may be associated with a particular defect or cluster of defects, or with a statistical fluctuation of the parameters of the material in some limited region of space. Electrons in these states have very low mobility at low temperatures.

The electronic states in the body of the valence and conduction bands are probably extended throughout the material. These states will have a much higher mobility than states in the quasigap, although their mobilities will be very much smaller than the mobilities of free carriers in crystalline semiconductors. The more or less sharp transition between

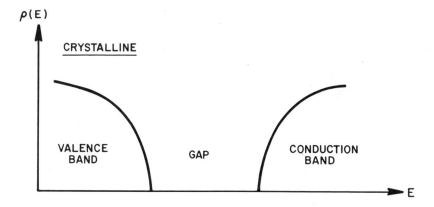

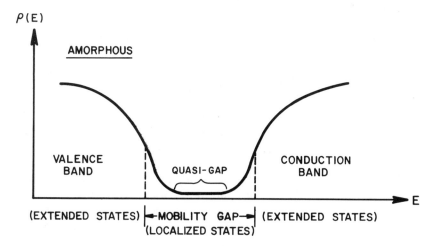

FIGURE 2   Schematic density of states for a crystalline and an amorphous semiconductor.

these states of relatively high mobility and the states of low mobility is called the mobility edge.

The general picture of amorphous semiconductors, presented above, is that underlying the theories developed by Mott, by Cohen, Fritzsche and Ovshinsky, and others (Mo 67b, Coh 69, 71b, Gu 63, Ba 64). Some of the more detailed models used to explain the properties of amorphous semiconductors will be discussed below.

We close this section concerning general concepts with a warning. Many terms such as *effective mass, n-type, p-type, intrinsic, extrinsic,* etc., have been borrowed from crystalline semiconductor theory and used to describe phenomena in the amorphous materials. It is important to remember, however, that the applicability of these ideas to the amorphous state is not always very well established and that, in any case, the precise meaning of the terms may be somewhat different than in the crystalline case.

OPTICAL PROPERTIES

The scientific and technological interest in amorphous semiconductors stems mainly from their electrical and optical properties. Usually, the optical properties are easier to interpret and we discuss them first. Optical absorption reflects essentially the density of states, more or less modified by the transition probabilities between the states. In crystals, it has been possible to obtain much information about the electronic and phonon state structures from the optical measurements, and similar methods have been applied to amorphous solids (Ta 70a, St 70a, St 70b, Ta 71).

Measurements

The precision of measurement is often severely limited by macroscopic inhomogeneities of the samples producing light scattering. The determination of the optical constants of films from transmission and reflectivity measurements meets with an additional difficulty. The optical constants are calculated from complicated equations which take into account interference effects. It often happens (in particular at low-absorption levels) that a meaningful determination of the optical constants requires not only a precision of measurement which is difficult to attain, but also an extremely uniform thickness and homogeneity of the film (Do 70). At high-absorption levels, the optical constants are determined from the measurement of the reflectivity and Kramers-Kronig analysis. As in crystals, one is concerned with the quality of the surface. In addition, there is the difficulty that the shape and position of the broad structure in the optical constants observed in amorphous materials is more sensitive to the arbitrary extension of the reflectivity data beyond the range of actual measurements than in crystals.

Principal Optical Absorption Band

A typical example of the difference in optical properties between crystalline and amorphous solids is shown in Figure 3. The principal absorption band of a-Ge is situated approximately in the same energy range as that of c-Ge, but it lacks the sharp structure characteristic of crystals. This is easily understood: with the loss of the long-range order, the k-vector ceases to be a good quantum number and is only partially or

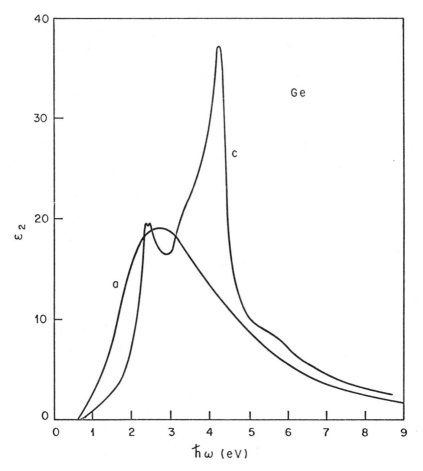

FIGURE 3   Principal optical absorption band of a-Ge (curve a) and c-Ge (curve c).

not at all conserved during optical transitions. In this case, the broad features of the spectra are determined by the convolutions of the band state densities rather than the joint band state densities whose singularities are responsible for the structure of crystalline spectra. The electroreflectance method, which is particularly sensitive to these singularities, appears to be less useful for the amorphous semiconductors. Recent work (Fi 71a) has shown that the previously reported electroreflectance in a-Ge is probably an artifact.

The simplest assumptions that the matrix element for the optical transitions is constant for the whole absorption band, and that the band state densities are little different from crystalline values have been checked for a-Ge and found to be unsatisfactory. Simple suggestions were tried to explain the shift of the principal absorption peak at 4.4 eV, dominating the spectrum of c-Ge, to 2.7 eV in a-Ge, based either on the change of the band state densities or the energy dependence of the matrix element (Br 69). The latter explanations appear to contradict the results on external electron photoemission (Sp 70) from amorphous Ge, which are consistent with the shift of the valence band state density toward the top of the band.

Only small differences in the principal absorption bands were observed in crystalline and amorphous forms of As_2S_3 and As_2Se_3 (Za 71), if we disregard the disappearance of the sharp structure in the amorphous form. In Se and Te, a large shift of the absorption edge toward higher energies is observed in amorphous as compared to trigonal forms. This is probably due to the disturbance of the interaction between the helical chains by disorder, and/or the appearance of the rings in the amorphous form.

Index of Refraction

From the available data, it appears that the electronic part of the index of refraction at low frequencies, $n(0)$, changes only moderately (at most ten percent) if the short-range order of the amorphous form is the same as in the corresponding crystalline form. This is the case for the network structures such as Ge, 3-5 compounds, As_2S_3, and others. This is consistent with the view that the average gap depends only on the short-range order. A large change of $n(0)$ is observed if the short-range order (character of the binding) changes as, e.g., in GeTe. GeTe crystallizes in a distorted rock-salt structure (coordination number 6). a-GeTe has a mixed coordination number of 4 and 2 for Ge and Te, respectively. The binding is predominantly covalent (Bet 70). The values of $n(0)$ are 6 and 3.3 in c-GeTe and a-GeTe (Br 70).

The twofold coordinated structures, Se and Te, exhibit a large change of $n(0)$ in accordance with the large shift of the band edge. The same effect is observed in some systems containing Te (such as $Ge_xTe_{1-x}X$), where X is some other constituent, from which tellurium-rich phases crystallize out during the amorphous–crystalline transition. These systems have a higher kinetic crystallization temperature than pure Te and are potentially useful materials for electro-optical memory devices.

Although $n(0)$ of a-Ge is close to that of c-Ge, it has been shown to be somewhat greater by 0 to $+10$ percent depending on the method of preparation, substrate temperature during deposition, and annealing (Th 71). It has been suggested that this is due to vacancies that disappear (or conglomerate into larger voids) during annealing. Electron-spin resonance observed in amorphous Ge has been associated with states at internal surfaces of voids (Br 69). The pressure dependence of $n(0)$ of amorphous Ge and Si is comparable to that observed in crystalline forms (Co 71).

Absorption Edge

In contrast to the principal absorption band, the shape and energy of the absorption edge often depends on the preparation and thermal history of the samples (a structure-sensitive property), in particular, in thin films. Figure 4 shows the dependence of the absorption constant of a-Ge in the region of the absorption edge on the temperature of annealing (Th 71). The pressure shifts of the edge in amorphous Ge and Si are different from the shifts of any of the gaps in the crystal (Co 71).

Similarly, as in crystalline semiconductors, the definition of the absorption edge energy, E_g^{opt}, requires some knowledge or at least some theoretical assumptions about the absorption processes in the region in question. It appears that in many amorphous semiconductors it is possible to distinguish between the high-energy part where the absorption constant varies as

$$\alpha \sim (\hbar\omega - E_g^{opt})^r/\hbar\omega \qquad \alpha \gtrsim 10^4 \text{ cm}^{-1}, \qquad (1)$$

the exponential part, where $\alpha \sim \exp(\hbar\omega/E_c)$, and the weak absorption tail, where $\alpha \sim \exp(\hbar\omega/E_t)$ with $E_t \gg E_c$ (for $\alpha < 1$ cm^{-1}). Here ω is the photon frequency and E_t, E_c are characteristic energies.

The high-energy part is often used for the definition of E_g^{opt}. For some materials, it is found close to E_g^{opt} in the corresponding crystal. In other materials, shifts in either direction have been observed. E_g^{opt} can be smaller or larger than the electrical gap determined from the temperature dependence of electric conductivity. This may indicate

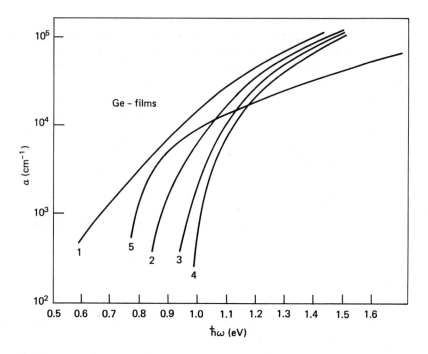

FIGURE 4   Dependence of the absorption edge of a-Ge on the temperature of annealing (Th 71). Curve 1: deposited at 20°C, nonannealed. Curves 2, 3, 4, 5: annealed at 200, 300, 400, 500°C, respectively. During the annealing at 500°C the sample crystallized.

that the Fermi-level is not necessarily in the middle of the gap. For the exponent, r, one often finds $r = 2$ (As$_2$S$_3$, As$_2$Se$_3$, GeTe, GdGeAs$_2$, Si, and others); sometimes $r = 3$ (in complicated chalcogenide glasses such as Si$_5$Ge$_{15}$As$_{25}$Te$_{55}$). In amorphous Se, the data indicate that $r = 1$ (Da 70).

Below $\alpha < 10^4$ cm^{-1}, the absorption edge is generally not as sharp as predicted by the formula $(\hbar\omega - E_g^{opt})^r$. In some cases, the broadening may be due to absorption associated with the states in the gap. This appears to be the case for amorphous Ge films which often show broad edges that can be sharpened by annealing (cf. Figure 4). Some methods of preparation give a sharp edge (Do 70). In many compound semiconductors, α, in the range of about 1 to 10^4 cm^{-1}, is proportional to $\exp(\hbar\omega/E_c)$ with the constant, E_c, in the range of 0.05 to 0.08 eV at room temperature. Lowering the temperature keeps E_c almost constant. At higher temperatures, E_c increases unless annealing processes take place, which may reduce E_c. This exponential part of the absorption edge behavior is similar to the Urbach tails observed in crystals, as discussed in the section on Models below.

Below the exponential absorption tail of compound amorphous semiconductors, one observes weak absorption tails with a considerably smaller slope (Ta 70b, Wo 71). The absorption in these tails depends very much on sample preparation and purity. It has been shown experimentally that this absorption is not an artifact produced by light scattering. This absorption tail has been associated with the states in the energy gap. From the available experimental data (Ta 70b), it appears that the optical absorption cross-section of these states is a few orders of magnitude smaller than the cross-section observed for localized states in crystals. Pure semiconducting glasses are highly transparent in the range below the absorption edge and above the lattice absorption bands.

Free carrier absorption is generally not observed even at high temperatures. Glasses are, for this reason, much more transparent at high temperatures in the infrared region than crystals with the same energy gap. However, in a relatively highly conducting glass, $Tl_2SeAs_2Te_3$, an absorption that may be ascribed to free carriers has been observed (Mi 71).

In the infrared spectrum of amorphous Ge, one observes a band at 0.23 eV, which may be interpreted as evidence for a peak in the distribution of the gap states. These states may arise from defects (Ta 70c).

In some chalcogenide glasses, a photoluminescence band has been observed at low temperatures (Ko 70, Fis 71) at energies well below the optical gap. This is not necessarily evidence for a peak in the localized state density in the gap as suggested originally, but can be explained on the basis of uniformly decreasing density tails and plausible assumptions about the competition of the radiative recombination process with the relaxation process.

ELECTRICAL PROPERTIES

Investigations of the electrical properties of amorphous semiconductors lean heavily on the procedures and analyses used in studies of crystalline semiconductors like Ge, Si, and the III-V compounds. However, exploration of amorphous semiconductors encounters many unusual aspects and difficulties (some of which are already apparent in work on low-mobility materials like oxides):

1. The preparation of well-defined samples for electrical measurements is a cumbersome and often impossible task. Besides foreign impurities and other traps, specimens contain uncontrolled compositional and positional inhomogeneities, which may give rise to charge accumulations and potential fluctuations.

2. The resistivity of an amorphous semiconductor or semi-insulator is often several orders of magnitude larger than that of its crystalline counterpart (Ow 70a).

3. Charge carrier concentrations are hard to obtain because of the difficulties in measuring and interpreting the Hall coefficient.

4. Mobilities are small and their magnitudes are difficult to evaluate. The determination of the Hall mobility depends on one's ability to measure the Hall coefficient. The drift mobility may be trap-limited and in any case is measurable only in sufficiently good samples.

5. The effects of impurities on electrical properties are hard to assess.

6. Many amorphous semiconductors show switching-type I-V curves and other nonlinear effects.

Experimental Techniques

Because of the inherent difficulties listed above, measurements on amorphous solids have the tendency to be ambiguous. Hence experiments are more convincing when they have been performed under a number of different conditions (a variety of substrates, sample thicknesses, substrate temperatures, quenching rates, electrical contacts, and, if possible, different sample geometries). In most cases, the *geometry of sample and electrodes* is much different from that used with crystalline specimens. Often samples can only be prepared in the form of thin films. Considering the high resistance of most amorphous semiconductors, it seems difficult to use anything else than the capacitor geometry. Frequently, this leads to uncertainties about the homogeneity of the applied electric field as a result of possible barrier layers at the electrodes. In dealing with relatively well-conducting amorphous materials, it should be possible to apply four electrodes and use the potentiometric method (at the same time being careful to avoid surface conduction). The capacitor geometry is also objectionable in determinations of the Seebeck coefficient. In this case, one measures the temperature difference between the electrodes, which will be quite different from that across the sample if sizable temperature drops occur at the metal–semiconductor interfaces.

Another aspect that has to be considered is the matter of *contacts*. In general, metal contacts on small-gap ($E_g < 1.5$ eV) amorphous semiconductors are believed to be ohmic and to have low resistance. On the other hand, problems arise with electrical contacts between metals and wide-gap amorphous materials (Fr 71a). In the latter case, both blocking and rectification have been observed.

Quite extensive studies have been made concerning contacts on amorphous Ge and Si (Bos 70). The conclusion reached is that the noble metals cause alloying and crystallization at temperatures as low as 100–200°C. In contrast, refractory metals do not cause such effects until much higher temperatures are reached.

D.C. Conductivity

Studies of a variety of amorphous semiconductors indicate that the temperature behavior of the conductivity depends on the type of bonding. For example, the purely covalently bonded amorphous Si and Ge show different electrical properties from amorphous Se and Te in which there is mixed covalent and van der Waals bonding. The chalcogenide glasses with their cross-linked network structures have again another kind of conductivity-temperature relation.

Most important is the fact that the temperature dependence of amorphous Si and Ge (ln σ vs. $1/T$) cannot be analyzed in terms of an exponential increase (not even over a limited temperature range). In other words, the conductivities of amorphous Si and Ge do *not* show a single activation energy (Br 71b, Cl 70). Recently it has been reported that the conductivities of amorphous thin films of Si, Ge, and also carbon below room temperature follow an exp (bT^{-1}) relationship rather accurately, in accordance with a tunneling mechanism suggested by Mott (Mo 69a). (See discussion further on in this section.)

The effect of impurities on the conductivity of amorphous Si and Ge differs considerably from that on crystalline samples of the same elements. Crystalline Si and Ge are strongly influenced by impurity concentrations as low as 10^{14} or 10^{15} cm^{-3}, while amorphous thin films do not seem to be affected before levels of 10^{18}–10^{19} cm^{-3} have been reached.

On the other hand, it has been pointed out in the discussion of the preparation of thin-film samples that amorphous semiconductors like Ge and Si are very susceptible to the temperature of the substrate, to temperature cycling (annealing) (Br 71b), and to the presence of even very small residual gas pressures (Wal 68).

The twofold coordinated semiconducting glasses like Se and Te show a simple exponential behavior of the conductivity over a wide temperature interval extending without any discontinuity into the liquid range (St 70a). The activation energies derived from these transport measurements agree roughly with one half of the optical energy gaps of the two materials measured on amorphous films. However, as remarked previously, this agreement may be coincidental. Amorphous Se and Te, which possess ring and chain structures, are much more sensitive to small impurity concentrations; amounts of the order of 10^{17} cm^{-3} appreciably alter the electron and hole mobilities (Fr 71a).

The conductivities of amorphous films of the chalcogenide alloys follow, in general, the same patterns of a well-defined exponential temperature dependence as discussed above for Se and Te. However, the conductivity changes more rapidly near the glass temperature, T_g. Cooling from temperatures around T_g, one finds in some amorphous films at lower temperatures different ln σ-vs.-$1/T$ curves depending on the rate of heating and cooling (Fr 71a). Rapid quenching tends to increase the activation energy at low temperatures. In glasses containing several constituents, there may be phase separation upon heating to temperatures above T_g; if this occurs, the effect on the electrical properties is considerable.

In summary, for the chalcogenide glasses, plots of log (conductivity) vs. reciprocal temperature often show a high-temperature segment which can be retraced in successive runs, and low-temperature branches which depend on previous history. In crystalline semiconductors, one would call these two parts "intrinsic" and "extrinsic." However, in the case of amorphous semiconductors these terms are not well defined. Perhaps it would be better to refer to the high- and low-temperature portions of the conductivity curves as structure-insensitive and structure-sensitive branches, respectively.

Hall Effect

As mentioned before, measurements of the Hall coefficient are difficult, and only very few have been performed. This situation is illustrated by the results of two experiments on amorphous Ge (Cl 67, Nw 68): The authors agree on the order of magnitude (10^{-2} cm^2/V-sec), but not on the sign. More reliable figures can be quoted for relatively well-conducting compounds such as As_2Te_3, Tl_2Se (Ko 60), and As_2SeTe_2 (Mal 67). In amorphous films of these materials, the Hall coefficient is negative; Hall mobilities are about 10^{-1} cm^2/V-sec and rather independent of temperature.

Drift Mobility

This parameter has been studied so far only in amorphous samples of Se (Tab 68), As_2Se_3 (Sch 70), and As_2S_3 (In —). Photocurrents were produced by very short light pulses. Transit times of the carriers were measured as a function of film thickness, wavelength, applied field strength, temperature, and light intensity. It was possible to separate drift mobilities and lifetimes; for Se the results indicate electron drift mobilities of the order of 6 to 8×10^{-3} cm^2/V-sec, while hole mobilities have surprisingly larger values of about 0.15 cm^2/V-sec.

Field Strength

In measuring electrical properties of amorphous semiconductors, one has to pay close attention to the strength of the applied field. Many of the samples used are thin films and hence the fields are large. Most semiconducting glasses show ohmic behavior up to fields of the order of 10^4 V/cm and increasingly larger deviations from linearity above this field. The latter can be traced back to two possible causes: electrode effects, and bulk effects (Fr 71a). The former result from injection at a metal–semiconductor contact. Although it is still somewhat of an open question whether injection can occur in most amorphous semiconductors, it is widely believed that carriers can be injected in films of chalcogenide glasses thinner than 1 μ, and that the switching in such films depends on the resulting space-charge effects. Hence the electrical current will be space-charge-limited and will depend on the applied voltage, because both the number of carriers and their drift velocity are proportional to the field. The presence of large numbers of deep traps will suppress the buildup of any space-charge-limited current.

Strongly nonohmic behavior has been observed also in cases where electrode effects do not play a role. The rapid increase of the current at high fields is then a bulk effect, which has been attributed to the lowering of a coulombic barrier by the action of an applied electric field, e.g., the Poole-Frenkel mechanism (Fr 71a, Jo 71, Bag 70).

A.C. Conductivity

A possibly fruitful technique for probing electrical properties of amorphous (and other low mobility) semiconductors is the measurement of the a.c. conductivity. A partial list of materials that have been explored by this method includes: Se, As_2Se_3, As_2Te_3, Te_2AsSi, and SiO (Fr 71b). Most measurements cover a frequency range from 10^2 to

10^8 Hz. At high frequencies ($\omega > 1$ MHz), the conductivity usually increases like ω^2 and is independent of temperature. Below 1 MHz, the frequency dependence of the a.c. conductivity is often somewhat less than linear: $\sigma \sim \omega^n$, where $0.7 < n < 1.0$. The latter behavior, which shows a small temperature dependence, has been compared with the a.c. response of Ge at very low temperatures (impurity range) (Po 61), where the ω and T dependence was interpreted as phonon-assisted hopping between trap sites.

One should realize that hopping is only one of many possible loss mechanisms that can occur in a semiconductor or semi-insulator. Besides charge carrier diffusion (corresponding to d.c. conductivity, which is ω-dependent), there are several other sources of dielectric losses (relaxation or polarization effects) such as inhomogeneities, dipole layers, surface barriers, ionic dipoles, etc. In most cases there is not one but a series of relaxation mechanisms, each with its own relaxation time.

Austin and Mott (Au 69) have derived an expression for the a.c. conductivity in amorphous semiconductors assuming the hopping mechanism. This will be discussed below in the part of this section devoted to models. Their expression contains the density of states at the Fermi level ρ_0. A.C. conductivity measurements have been used to obtain values of ρ_0 for various materials. The result, however, is somewhat suspect because of the ambiguity of interpreting the a.c. loss and the uncertainty about the magnitude of several other parameters in the hopping mechanism.

It may be helpful to study very simple glasses, where one can isolate, hopefully, a single Debye peak. Studying such a peak under a variety of different external circumstances (electric field, photoexcitation, different dopings, etc.) might provide the key to an understanding of the distortions and reorientations that take place in the glass. One should realize, however, that any conclusions reached concerning the mechanism responsible for a.c. conduction behavior do not necessarily extrapolate to the d.c. case.

Photoconductivity

The exploration of photoconductivity in amorphous semiconductors has produced a host of interesting relations between the photocurrent and such experimental variables as wavelength, intensity and duration of the exciting light, temperature of the sample, or applied field. Explanations of these effects are usually in terms of phenomenological models.

Photoconductive effects of widely varying magnitudes have been observed in a-Ge and Si (Gr 67, Fi 71b), in a-Se, in a-GeTe, and in chalcogenide glasses (Jo 71). In Se, the photoconductive threshold appears 0.6 eV above the absorption edge. The smaller value of the strong absorption is identified with a localized exciton of the Se_8-molecules, while the larger threshold for the photoconductive process is associated with transitions between one-electron band states (Lu 70).

As_2Se_3 and other chalcogenide glasses do not show a well-defined photoconductivity threshold. In these materials, carriers seem to be excited from one-electron states extending well into the forbidden energy gap. Experiments have concentrated on the dependence of the photocurrent on light intensity and temperature, and on transport effects. In order to explain the results of these measurements, several investigators (We 70, Ho 70) have assumed recombination or cross-section edges located between the midgap and the mobility edges. The recombination edge, E_K (or cross-section edge), is defined as that energy where recombination occurs at the same rate as thermalization.

Two other photoeffects have been observed in chalcogenide glasses (Fa 70). The first known as steady-state photoconductivity: Illumination of a thin film at nitrogen temperature produces an excess current, which decays extremely slowly after cessation of the exciting light. The other effect is the generation of a thermally stimulated current after brief exposure to pair-producing radiation at 77°K. These phenomena may contribute to our knowledge of electronic states (traps and tail-states) within the pseudogap.

Seebeck Coefficient

As mentioned before, data concerning this parameter are rather uncertain because of the difficulties involved in the measurement (especially in the case of thin films). If one takes the results at face value, it appears that qualitatively the temperature dependence of the Seebeck coefficient is not very different from that of crystalline semiconductors (St 70a, Ow 70b). The Seebeck effect is a first-order effect in nondegenerate semiconductors; hence an analysis of the effect requires a rather exact knowledge of the density of states, the location of the Fermi level, and the nature of the scattering mechanism. Unfortunately, such information is practically unavailable.

For a series of chalcogenide glasses, the Seebeck coefficient is positive, indicating hole conduction, while Hall coefficient measurements on the same samples show negative values, implying that conduction is by electrons. In a complicated system, however, these two measurements need not be related in a simple way.

Piezo- and Magneto-Resistance

The piezoresistive and magnetoresistive effects have been measured for a few amorphous semiconductors (St 70a). In all cases, the effects are very small ($\leq 1\%$) and negative. From the experimental point of view, some warnings should be expressed. It is not clear that external stress is meaningful when the amorphous sample is already heavily strained. A negative magnetoresistance could be the result of inhomogeneities or internal barriers (Her 60). Furthermore, one should realize that transport properties depend on a host of material parameters, which makes any analysis of experimental results extremely difficult.

Very recent measurements of electron tunneling into a-Si (Sau 71) show promise for obtaining information concerning the density of localized states.

MAGNETIC PROPERTIES

The magnetic susceptibility of glasses is generally diamagnetic and constant at high temperatures and exhibits a superimposed Curie-term at low temperatures (Ta 70c). The glasses are always more diamagnetic than the corresponding crystals. A satisfactory theoretical explanation of this fact is not available. From the Curie term, the concentration of free spins can be estimated. One assumes that they correspond to highly localized states in the gap because such states may be expected to be singly occupied. Their concentration in some chalcogenide glasses was found to be of the order 10^{18} cm^{-3}. By measuring the same sample of As_2S_3 in its crystalline and amorphous forms, it was shown that these states are due to disorder (Ta 70b).

It has been possible to measure electron-spin resonance in films of amorphous Si, Ge, and SiC. The spin concentration in nonannealed samples was found to be of the order 10^{20} cm^{-3} and diminished with annealing (Br 69). The resonance was ascribed to states on internal surfaces associated with voids. Recently ESR has been observed also in chalcogenide glasses (Smit 71).

Faraday rotation has been measured in amorphous Se at the absorption edge. The results were found to be interpretable in terms of nondirect transitions (Mort 71).

LATTICE VIBRATIONS

Because of the absence of k-vector conservation, the phonon spectra, both infrared (Ta 70c) and Raman (Sm 71), reflect essentially the phonon state densities. In amorphous Ge, Si, and 3-5 compounds, it appears from the experimental data that these densities are similar to those observed or calculated in crystals. In a-Se and compound semi-conductors, reststrahlen bands are observed also in the amorphous state; however, the bands are broadened and the weaker ones often disappear.

The relaxation of the k-vector conservation often leads to disorder-induced bands, such as the appearance of one optical phonon absorption band in a-Ge (Ta 70c), which is forbidden in c-Ge. Another example is a broad absorption band observed between 1 and 300 cm^{-1} in oxide glasses, which has been tentatively ascribed to single acoustical phonon absorption (Am 69); these bands have not yet been studied in amorphous semiconductors.

Inelastic scattering of neutrons on polycrystalline and amorphous Se has shown little difference between both spectra (Kot 67). The authors conclude from this that the phonon state density is determined predominantly by the short-range order.

Localized vibrations of impurities and defects have been observed in semiconducting glasses. In a-Ge, bands probably associated with vibrations of oxygen-defect complexes have been reported (Ta 70c). The presence of oxygen may severely limit the infrared transparence of chalcogenide glasses because of vibrations due to oxygen bonds (Va 70).

In connection with long wavelength acoustical phonon excitations, crystalline as well as noncrystalline solids may be regarded approximately as elastic continua. The use of glasses in ultrasonic applications is there-fore often convenient. Recently, some semiconducting glasses (such as $Ge_{30}As_5S_{65}$) have been shown to exhibit very low acoustical losses (Kra 70). This makes them attractive materials for acoustic delay lines and also for acousto-optic devices.

Thermal conductivity of glasses is due to phonons since the electronic contribution is generally negligible. At room temperature the thermal conductivity is much smaller than in crystals. This is explained by the very small mean-free-path of thermal phonons, of the order of inter-atomic distances. However, at very low temperatures much longer mean-free-paths have been observed (Ze 71).

The specific heat at very low temperatures in a number of semi-conductor and oxide glasses shows a deviation from the Debye law. Between 0.1°K and 3°K it has been observed to vary as $AT + BT^3$ (Ste 71). This departure from the expected behavior may possibly be generally characteristic of the amorphous state and is sufficiently re-markable to warrant further investigation. Its origin is not yet understood.

MODELS

Density of States and Optical Properties

Various models have been used to apply the experience with the band structure and optical properties of crystalline semiconductors to the problem of obtaining a quantitative understanding of their amorphous counterparts. Comparison with the crystalline case is most likely to be useful for the extended states of the amorphous material.

For a substance such as Se, which occurs in several crystalline forms, one approach has been to approximate the density of states and optical properties of the amorphous form by some weighted average over the crystalline forms (St 69). A related approach is the work of Rudge and Ortenberger (Ru 71) who used the OPW method to calculate the band structure and optical properties of *hypothetical* crystals of Ge in the 2H (wurtzite) and 4H structures, as well as for the actual cubic (diamond) structure. They then compared the observed dielectric function of amorphous Ge with an average for the various crystals, achieving at least partial success in explaining the differences between the observed spectra of a- and c-Ge shown in Figure 3.

Henderson (Hen 71) has constructed a periodic tetrahedrally co-ordinated lattice having 64 atoms per unit cell with the local atomic arrangements simulating an amorphous array, whose radial distribution function is in reasonable agreement with the electron diffraction results for a-Ge (see Section II). If it is possible to calculate the energy levels of such a system, we should gain considerable insight into the nature of the electronic states in the bulk of the conduction and valence bands in a-Ge, and we might hope to reproduce with considerable accuracy the density of states and the optical absorption spectrum in the region of band-to-band transitions.

Another class of models that has been used to understand amorphous systems starts from the wave functions or pseudopotential appropriate to the periodic system, and then introduces the effects of disorder as a more or less ad hoc scattering process. Examples of this approach are the calculations of Kramer (Kram 70) on a-Se and the work of Brust (Bru 69) on a-Ge.

A rather different approach to understanding amorphous semiconductors was taken by McGill and Klima (Mc 70). They calculate the "reaction matrix" for a cluster of eight tetrahedrally coordinated carbon atoms in a staggered and in an eclipsed configuration, using a muffin-tin approximation for the atomic potentials. Multiple scattering theory then yields a density of states for a hypothetical amorphous diamond, the system being treated as a gas of randomly distributed eight-atom-molecular scatterers.

A qualitative understanding of the optical properties of semiconductors, amorphous and crystalline, can be obtained from the Penn model of an isotropic semiconductor (Pen 62). In this model, the electronic properties are characterized by two parameters: the density of valence electrons, and an average energy gap, Δ, which corresponds to the average energy gap along the faces of the Jones zone in the crystalline case (Hei 69). Perhaps more sophisticated theories based on this model will be able to deal with detailed differences between amorphous and crystalline materials. Phillips' theory of chemical bonding in semiconductors, which emphasizes the dielectric constant of the materials, the density of electrons, and the electronegativity of the atoms, has had remarkable success in systematizing optical properties, heats of formation, and elastic constants of the tetrahedrally coordinated crystalline semiconductors (Ph 70a). This approach may also prove useful for the amorphous case.

Attempts to understand the form of the optical absorption near and below the fundamental absorption edge have been generally less concerned with explaining the features of particular materials than with elucidating the widespread occurrence of certain forms. The common occurrence of an imaginary part of the dielectric constant proportional to $(\hbar\omega - E_g^{opt})^2$ [cf. Eq (1)], just above the fundamental absorption edge, can be explained as either an indirect or a non-direct transition between a valence band and conduction band having the usual square-root dependence of the density of states at the band edges (Ta 65). It has been remarked that the magnitude of the absorption edge in amorphous Si, relative to the indirect edge in crystalline Si, can be simply explained assuming a mean-free-path of 10A in the amorphous form versus 350A in the crystal (Eh 70). Davis and Mott (Da 70), however, explain the absorption just above the fundamental edge in terms of transitions from extended states in the valence band to localized states at the edge of the conduction band, or vice versa. The quadratic dependence then arises from the convolution of a finite density of extended states at the mobility edge, and a density of localized states which they claim to be roughly a linear function of energy. Cases where the quadratic law does not hold (e.g., Se) are explained in terms of peculiarities in the densities of states for these substances.

Localized states in the pseudogap pose rather different problems than the states in the bands. For the localized states, the kinds of defects or fluctuations responsible for the binding of the states must first be identified, the binding energy associated with particular kinds of defects must be calculated, and the expected probability for finding the various kinds of defects must be understood on a statistical basis.

If a localized state is tightly bound to a single defect such as a foreign impurity atom, a vacancy, etc., it may be possible to estimate the binding energy or range of binding energies possible for the state by considering the experimentally observed binding energy of a similar defect in a crystalline semiconductor. To the best of our knowledge, good first-principle calculations of the binding energy of a deep trap in a crystalline semiconductor have never been done. These should be possible with presently available techniques. The study of defect states in a quasi-amorphous lattice, such as the Henderson model mentioned above, might also be very enlightening. Estimates of the numbers of various kinds of defects are difficult to make, but experimental measurements of various kinds, such as density, magnetic resonance and susceptibility, Raman scattering, diffraction, etc., may be useful in deciding between different statistical models.

For localized states bound to a large cluster of defects, or to a multi-atomic statistical fluctuation in the density, composition, or short-range order of the amorphous structure, detailed microscopic calculations of wave functions do not seem at all promising. Here, one has been forced to use such simplifications as the effective-mass approximation, deformation potentials, etc. For various simple assumptions about the probability distribution of the relevant fluctuations, it is possible to estimate the density of states and perhaps the optical properties, using methods similar to those developed to investigate the effects of fluctuations in the impurity distribution in crystalline semiconductors (Li 63, Bon 62, Ka 63, Mor 65, Ha 66, Zit 66). Using this kind of approach, Stern has been able to fit observations of the optical absorption in poorly annealed samples of amorphous Si, in the region below and just above the fundamental edge (Ster 71a). Stern assumes a fluctuating deformation potential, obeying a Gaussian distribution with a short correlation length ($\sim 6A$), and with a root-mean-square potential that depends on the annealing history of the sample. Using his wave functions and density of states, in conjunction with Mott's theory of hopping conductivity (described below), Stern has also been able to explain the observed d.c. conductivity in amorphous Si (Ster 71b). Unfortunately, Stern's assumptions do not have a simple physical interpretation. Moreover, it is very difficult to test the assumptions underlying this kind of description, since it is entirely possible that similar transport and optical properties could result from very different models for the fluctuation statistics.

Various pieces of experimental evidence have been cited recently to support the idea that the most important fluctuations responsible for states in the quasigap result from *long-wavelength* ($\sim 100A$) electric fields in the sample (Fr 71b, Mo 67a, b, Ow 67, Boe 70a, Bag 70, Ta 70d). The effect of the resulting electrostatic potential would be to shift the position of the valence and conduction band edges locally, relative to the Fermi level. If the fluctuations in the potential are large enough, the Fermi level may sometimes be above the bottom of the conduction band and sometimes below the top of the valence band. This would lead to a finite density of states at the Fermi level, even in the absence of deep trap states bound to point defects (see Figure 5). A model with long-wavelength potential fluctuations might account for the relatively large density of states inferred from susceptibility measurements (Wo 71) and from various electrical measurements in the chalcogenide glasses (Fr 71b), while at the same time explaining the very low-optical absorptions inside the quasigap, and the long recombination times for trapped carriers seen in these materials. The long recombination times result from the large spatial separation between the regions where electrons and holes will respectively accumulate. The absence of substantial optical absorption well below the fundamental edge is accounted for in that a potential fluctuation causing a depression of the

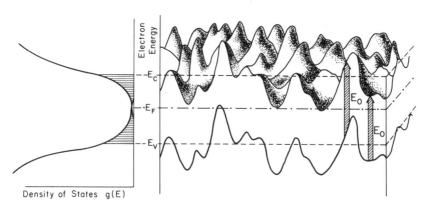

FIGURE 5 The heterogeneous model. [Reproduced by permission from (Fr 71).] The right-hand side shows the edges of the valence and conduction bands modified by long-wavelength electrostatic potential fluctuation. Optical transitions take place with an energy gap E_0 as shown The left-hand side shows the density of states. The region of localized states lies between the mobility edges E_c and E_v. Short-range potential fluctuations that may give rise to additional localized states are not shown here. Any nonelectrostatic long-wavelength fluctuations that cause a variation of E_0 are omitted from this figure.

conduction band edge also causes a depression of the valence band edge, and vice versa. The large spatial extent of the fluctuations here serves to prevent absorption via photon-induced tunneling between a valence band state and a conduction band state, localized in regions of negative and positive electrostatic potential respectively. A model of this type, in which the spatial range of fluctuations is sufficiently large to prevent overlap between bound states for electrons and holes, will be referred to as a *heterogeneous model*.

A possible source of the hypothesized electrostatic potential fluctuations is a random distribution of charged defects, such as atoms which do not locally satisfy their valence requirements. Long-wavelength statistical fluctuations in the position of the valence and conduction band edges could also arise from composition fluctuations in systems that are close to a phase separation, or as a result of fluctuations in the conditions of deposition of the sample. Such composition fluctuations would not normally preserve the relatively sharp threshold for optical absorption, however. It remains to be demonstrated whether the above-mentioned fluctuations do in fact occur with the required statistical distribution and spatial variation to explain the observations in the chalcogenides. (There is some evidence to support a heterogeneous model for amorphous Ge and Si, as well as in the chalcogenides, but the experimental case is weaker in the former case [Fr 71b].)

As already pointed out in connection with optical properties, the exponential tail of the optical absorption below the fundamental edge of well-annealed amorphous semiconductors bears a strong resemblance to the Urbach tails observed in crystalline materials, from partially ionic semiconductors to alkali halides. Optical absorption tails can be produced, for example, by interaction with phonons, from the Franz-Keldysh effect in the electric fields of random impurities and defects, and by frozen-in strains. There have been many attempts to calculate the frequency- and temperature-dependence of the absorption tail based on these mechanisms (To 59, Ho 61, Kei 66, Ha 65, Re 63). There are difficulties with all of these attempts, however (Hop 68). Recently, Dow and Redfield (Dow 70) have obtained an exponential dependence of the optical absorption from numerical solution of a model of an interacting electron-hole pair in the presence of random electric fields caused by charged impurities in semiconductors, or by optical phonons in ionic insulators. Davis and Mott (Da 70) attribute the temperature-dependence of the steepness of the exponential edge, commonly observed in amorphous semiconductors, to a temperature-dependence of the dielectric constant entering the Dow-Redfield calculation. Final resolution of the Urbach-tail problem is still awaited.

Phillips has emphasized the importance of a self-consistent consideration of the displacements of atomic positions in understanding the binding energy of impurity states in crystalline semiconductors (Ph 70a). In view of the relative ease with which atoms in an amorphous structure can rearrange themselves, these effects may be even more important in determining the nature of localized states in the amorphous semiconductors, and may therefore be an essential ingredient of a correct theory of these states (Ph 70b).

Transport Properties

Two kinds of mechanisms have been proposed for the d.c. electrical conductivity of an amorphous semiconductor: a "free-carrier" contribution resulting from electrons or holes thermally excited into the high mobility extended states, and a "hopping" contribution resulting from phonon-assisted tunneling between localized states of random energy, not too far from the Fermi level. The free-carrier contribution has an activation energy which is the separation between the Fermi level and the nearest mobility edge. As noted earlier, this activation energy need not be equal to one-half the optical gap, as in the crystalline case, since the width of the mobility gap need not coincide precisely with the width of the optical gap, and since the Fermi level need not lie precisely in the middle of the quasigap if there is a finite density of states there. For the hopping conductivity, Mott has proposed a form (Mo 69a):

$$\ln \sigma \propto - (\alpha^3/\rho_0 kT)^{\frac{1}{4}}, \tag{2}$$

where ρ_0 is the density of localized states at the Fermi level, α^{-1} is the length for exponential decay of the states, and the variation of these quantities with energy is neglected. The unusual temperature dependence arises from the existence of a continuum of possible activation energies for hopping between two localized states of random distance from the Fermi energy. At low temperatures, pairs with low-activation energy become important. The activation energy cannot be made too small, however, because the average spatial separation between sites acceptably close to the Fermi energy would then be large, and this would make the tunneling matrix element small. The hopping conductivity can have forms other than (2), including a simple activation-energy form, when the assumption of constant density of states is not valid (Da 70). It is not clear why amorphous Ge, Si, and C show the form (2), while other amorphous materials show activation-energy forms for their conductivities. It is also possible that some unknown mechanism, quite different from the ones considered, is responsible for one or both of the observed conductivity behaviors. Other functional fits to the experimental data must also be considered (Ch 70).

The mobility of carriers in the extended states of the valence or conduction band is difficult to measure directly, since observed drift mobilities are probably strongly trap-limited. Theoretical estimates of the lower limit to the mobility of an electron in an extended state have generally ranged from 5 to 100 cm^2/V-sec at room temperature (Io 60). (These estimates are obtained by equating the mean-free-path to the interatomic spacing, or to the thermal de Broglie wavelength, respectively.) In c-NiO, however, an observed Hall mobility of 1 cm^2/V-sec has been attributed to extended-state conduction (Bosm 66). Davis and Mott (Da 70) interpret transport measurements in the chalcogenide materials using a band mobility of the order of 10–50 cm^2/V-sec.

A.C. Conductivity

When the d.c. conductivity of a material is small, it is often the case that the a.c. conductivity is large compared to the d.c., even at relatively low frequencies. Austin and Mott (Au 69) obtained the following result for the a.c. conductivity arising from phonon-assisted hopping between localized states:

$$\sigma(\omega) = \frac{\pi}{3}\rho_0{}^2 k T e^2 \alpha^{-5}\omega[\ln(\nu_p/\omega)]^4, \tag{3}$$

where ν_p is a typical phonon frequency, and the remaining symbols are the same as in the d.c. expression (2). This result has already been mentioned in connection with the discussion of electrical properties. As noted there, Equation (3), which is based on the results of Lax, Pollak, and Geballe (Po 61) for impurity bands in crystalline semiconductors, is indistinguishable from a frequency-dependence of the form ω^s, where s is about 0.8. The a.c. conductivities of a number of materials have been fit to this expression.

In addition to the mechanism included in (3), there will be a contribution to σ (ω) from simple *photon*-assisted tunneling between localized states (no phonon necessary) which will be present even at $0°K$. The associated frequency-dependence has been predicted to be of the form $\omega^2(\ln \omega)^4$ (Mo 69a). Such a contribution may have been observed in several materials (Ar 68, Cha 69).

Basic Theory of Disordered Systems

Many of the properties of amorphous semiconductors should be characteristic of disordered systems in general, and indeed many of the important ideas in this field have originated from the study of simpler disordered systems. Conversely, efforts spent in understanding amorphous systems may turn out to be useful in the study of other disordered systems. Among the systems whose electronic properties have attracted attention in recent years are systems with *compositional disorder,* such as metallic substitutional alloys and impurity bands in crystalline semiconductors, as well as systems showing *structural disorder,* such as liquids and gases. The need to consider effects of strong multiple scattering arises not only for electronic states in disordered structures, but also for phonon propagation and for the propagation of electromagnetic waves through sufficiently irregular media. Furthermore, many aspects of disordered systems seem to be related to unsolved problems in the theory of turbulence in fluids, the theory of second-order phase transitions, and the quantum mechanical many-body problem in general.

Conceptually, the amorphous semiconductor is one of the most difficult cases to treat. For a substitutional alloy, it is usually possible to understand the electronic states in terms of a periodic model potential, which is in some sense intermediate between the potentials or pseudo-potentials of the individual constituents. Ideally, the scattering due to the difference between the actual and model potentials is small.

For liquid metals, it is often possible to treat the electrons as nearly free, the scattering of the electrons from the ions now being treated as a weak perturbation. In this case, scattering contributions from the different Fourier components of the ionic pseudopotential are additive, and only the structure factor, or pair distribution function of the liquid need be specified. In amorphous or liquid semiconductors, however, particularly for states near the band edges, it is clear that multiple scattering of the electrons is very important, and the short-range order, involving correlations of three or more atoms, plays an essential role.

Many approximate techniques, of varying degrees of sophistication based on one form or another of perturbation theory, have been developed to study extended states in disordered systems, when the scattering is not too strong. These techniques—which include the multiple scattering theory of Lax (Lax 51); the Green's function approach of Klauder (Kla 61); the various Green's function theories for liquid metals developed by Edwards (Edw 67), Ziman (Zi 66), Anderson and McMillan (An 67), and Schwartz and Ehrenreich (Schw 71); the "virtual crystal approximation" (Nor 31), "average t-matrix approximation" (So 66), and the "coherent potential approximation" (So 67) for binary alloys; the cluster expansions used in alloys (Yo 68, Ai 69) and in liquid metals (Cy 66)—are valid when the mean-free-path is greater than the characteristic wavelength of the electrons. Transport properties in this regime may be handled by means of a Boltzmann equation or by Green's function approximations based on the Kubo formulas, such as in the calculation of Velicky (Ve 69). The situation where the mean-free-path is comparable to the characteristic wavelength of the electrons is much more difficult to handle, however, even for the case of a purely random distribution of scatterers, and quantitatively accurate methods are not known. Nonetheless, various of the Green's function methods have been applied to simple models in this regime in order to estimate such properties as the density of states, the mobilities, and the position of the mobility edge (Ki 70, Ec 70a). These methods may someday find extensions to the case of amorphous semiconductors.

The original suggestion of a sharp transition between localized and extended states in a disordered system was made by Anderson, using a model appropriate to spin-diffusion in disordered structures, or for electron transport in a tight-binding impurity band (An 58). Much of the subsequent discussion of this transition has been based on this kind of tight-binding model with random site energies (Zi 69, Ec 70b, Tho 70). Nonetheless, the crucial conclusion is believed applicable to many systems including amorphous semiconductors (Mo 67b, Coh 69). This is the conclusion that the one-electron levels of an infinite, macroscopically homogeneous, disordered system fall into two distinct classes—extended and localized. Furthermore, at any given energy, all of the states belong to the same class, except at a few isolated energies (mobility edges) where there is a boundary between a region of localized and a region of extended states. If the Fermi level falls in a region of localized states, then the d.c. conductivity will be zero at $0°K$; otherwise, the conductivity remains finite at low temperatures. The inclusion of electron–electron and electron–phonon interactions should not change this situation for the ground state at $0°K$ (An 70), although the interactions must blur the distinction between localized and nonlocalized states at finite temperatures, or for high-energy states even at absolute zero.

The energy-dependence of the mobility of an electron just above the mobility edge is likely also to be independent of the details of the disordered system. Several authors have attempted to investigate this behavior theoretically using simple models (Da 70, Eg 70). The close relationship between the localized–delocalized transition and the transition in percolation problems has been emphasized. The concepts of percolation theory may prove useful in understanding a number of aspects of the transport process (An 58, Eg 70, Amb 71).

Some insight into the properties of random systems, and checks on the validity of various approximation methods, can be obtained from the study of various exactly soluble models. For example, for a large class of one-dimensional random systems, one can calculate with arbitrary accuracy the density of states and the average one-particle Green's functions (i.e., the spectral density as a function of wavevector, $\vec{k}$, as well as of energy, E) (Ha 68, Hor 68). Similar exact results can be obtained for a three-dimensional, tight-binding model with a special distribution (Lorentzian) for the random site energies (Li 69). Rigorous theorems have been proved concerning the existence of energy gaps in various simple models of disordered systems in both one and three dimensions (Hor 68, Ki 70), including a model with some resemblance to a tetrahedrally coordinated, amorphous semiconductor (Wea 71, Hei —). Theorems that exactly give the low-order moments of the density of states for various tight-binding models have also been exploited (Mon 42, Ka 62, Brin 70).

An interesting feature of the exact results for the "Lorentzian" model (Li 69), which was rigorously extended to a wide class of models by Edwards and Thouless (Edw 71), is the fact that the density of states and the averaged one-particle Green's functions are analytic functions of the energy on the real axis, even in the regions where the localized–delocalized transition is supposed to occur. Thus the mobility edge is not reflected in these properties of the system. In one-dimensional random systems, all the one-electron states are believed to be localized, and the d.c. conductivity always is zero, if the model has no electron–phonon or electron–electron interactions (Bor 63, Mo 61, Ha 68, Lan 70).

Let us remark in conclusion, that although the theoretical studies described above have been useful in clarifying concepts invoked in the description of amorphous semiconductors, they are still rather far from providing the basis for a quantitative understanding of amorphous semiconductors. For instance, the question of whether there is a mathematically precise transition between localized states and nonlocalized states at $0°K$, which has attracted so much theoretical interest, is of less practical interest than, say, the question of the *location* of the mobility edge in a real amorphous material. (Of course, experimentally observable quantities would be little affected if, in fact, the transition between localized and nonlocalized states were somewhat ill-defined, and there were simply a rapid change in the magnitude of the mobility over some narrow range of energy.) It would seem that only limited progress can come from the study of abstract models, and that most of the effort will have to be spent in relating the electronic properties to the structure of real amorphous materials. The construction and analysis of realistic models is a difficult task, but is a challenge that must be faced if real progress is to be made. The continued evolution of computer technology and computational techniques is a factor that greatly enhances the eventual prospects for analysis of realistic models.

VI

Device Physics

SCOPE OF APPLICATIONS

Amorphous semiconductors lend themselves to a wide array of possible uses. A few of these, which have particularly stimulated fundamental work in recent years, will be singled out for detailed discussions. These will include:

(a) The use of inorganic amorphous semiconductors in electrophotography, i.e., document copying through the use of an electrostatic image of the material to be copied. This image is produced, after corona charging, through a selective discharge by photoconductivity, and the electrostatic image is used in turn to control the deposition of charged pigment particles.

(b) The use of selective crystallization (or phase separation) in amorphous films, through incident light, in image handling methods, such as photography and electrophotography.

(c) The use of the same sort of selective crystallization (or phase separation) induced by a laser beam to write digital information. The laser beam can also produce local melting and quenching and thus can restore the amorphous material to its original form. The information

59

thus written and/or erased can be read out through the significant change in optical properties between the amorphous and crystalline state.

(d) The use of amorphous chalcogenides in a negative resistance device. Any nonlinear resistance in which current and voltage are not monotonically increasing functions of each other is a negative resistance device. If the current–voltage characteristic gives voltage as a single-valued function of current, we speak of a current-controlled device. If, in the I vs. V plot, V is the abscissa, this characteristic will typically be "S" shaped, and is therefore called an "S" type negative resistance. If current is a single-valued function of voltage, a voltage-controlled device, sometimes called an "N" type negative resistance, is involved. The devices of principal concern here are current-controlled devices of the type illustrated in Figure 6a.

(e) The use of somewhat similar devices (particularly compositions containing 81 percent Te, 15 percent Ge, and 2 percent each of two minor constituents) in which the resistance at zero voltage is history sensitive. These can be left in a high- or low-resistance state (Dew 62) as illustrated in Figure 6b and used in random access computer memories.

There are other applications that make less use of the properties emphasized in the preceding sections of this report and where the device work has been less intimately coupled to the attempt to understand the kinetics of the materials.

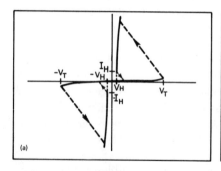 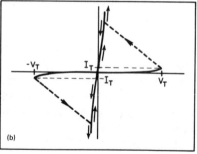

FIGURE 6 (a) Current–voltage characteristic of a typical current-controlled negative-resistance chalcogenide device. The dotted lines are primarily significant for their end points, showing the transitions made with a particular value of series resistance, as the total voltage applied to the series combination is varied. See, for example, (Ov 68a).

(b) Current–voltage characteristic of a memory device, showing bistability at $V = I = 0$.

Even though they fall outside the scope of this report, it may, nevertheless, be of help to list some of the more diversified applications. An extended discussion of proposed applications for amorphous *oxide* films has been given by Dearnaley *et al.* (Dea 71). This review discusses cold cathode electron emitters, display panels, electroluminescent devices, oxide-coated cathodes, triodes, microphones, and others. The use of glasses, including chalcogenide glasses, as infrared windows was the subject of an earlier Materials Advisory Board Report (MAB-68). Applications to acoustic delay lines and to infrared acousto-optic devices have already been mentioned in the earlier section on optical properties. Other optical applications include the bonding of germanium prisms to gallium arsenide film. The use of amorphous semiconductors in vidicons, i.e., as photoconductors in television camera tubes has been investigated. The May 1951 issue of the *RCA Review,* in a series of articles (Ros 51), discusses this role of photoconductors, including Sb_2S_3, and Se in particular. Sb_2S_3 is used in commercial cameras for studio transmission from motion pictures. Electrolytic capacitors utilize Al_2O_3 in low-cost applications, and Ta_2O_5 in more demanding high-capacitance devices. Bell Laboratories has, in fact, developed a complete tantalum-based technology for integrated passive circuits (McL 64, 66), and this technology is widely used and manufactured within the Bell System. Tantalum nitride and tantalum with, and without, interstitial oxygen content are used as the resistive materials, Ta_2O_5 as a dielectric. The use of SiO_2 as an insulator and passivator in the silicon technology is well established. Particularly demanding requirements are imposed on SiO_2 in the Insulated Gate Field Effect Transistor. In a recent modification of this transistor, the MNOS (Metal–Nitride–Oxide–Silicon) device, the insulator consists instead of a thin layer of SiO_2 in contact with the silicon, and a much thicker layer of Si_3N_4 on top of that. In this device (which will be further discussed in the section on technological comparisons), tunneling through the SiO_2 layer is an intentional part of the device's operation. Another application of As–S glasses relates to the passivation and encapsulation of devices, where there is a need to keep the previously manufactured device at low temperatures (Fla 60).

The rich variety of phenomena available in the oxide glasses (Dea 71, Si 70) includes switching devices. Oxide glasses exhibit both current-controlled and voltage-controlled negative resistance. The larger gap materials tend to be accompanied by voltage-controlled negative resistance. Sometimes both types of switching can be found in one material (e.g., Nb_2O_5). One oxide device particularly worth noting is an Nb_2O_5 memory device, which will be briefly discussed in the next chapter on technological setting. Other oxide devices, e.g., copper bearing glasses can also be switched repetitively at useful rates (1 MHz) and 10^8 times before failure (Dra 69). Oxide devices do not seem to depend on phase changes and phase separation in the manner characteristic of the chalcogenide memory devices, and we will return to this point in the section on technological comparisons. The oxides have been studied less intensively in recent years than the chalcogenides and offer an even less coordinated array of phenomena and models than is found in the chalcogenides. The proposals for oxide applications are also at an earlier stage of development, typified by the characterization of oxide switches (Dea 71):

The chief objection lies in the fact that the characteristics are not reproducible, from one operation to the next, or over a useful life. Most structures cannot be used in atmospheric conditions, and require encapsulation which adds substantially to the cost, but even when so protected the switching threshold is variable and there is a lack of stability below the threshold voltage. Over successive operations the devices frequently deteriorate and show evidence of local overheating: they may eventually persist in either the high or low impedance state. The switches are also rather noisy.

Chalcogenide glasses, which have benefited from more development activity, are not as much beset by these difficulties. For example, the variability in switching voltage for a given delay time is small, well above threshold (Lee 71).

In reading the following discussion of the key device areas, it is well to remember that there is proprietary information in this field, not available to this committee. We may be posing some questions that are in fact already answered. Furthermore, some of the available knowledge stems from private communications rather than published data.

IMAGE HANDLING AND OPTICAL MEMORIES

Electrophotography

Electrophotography, as currently used in document-copying machines, is undoubtedly the major existing application for amorphous chalcogenides. The particular form of electrophotography historically associated with the Xerox Corporation utilizes Se and As–Se compositions as the photoconductive element of the system. These materials have high resistivities in the dark, combined with reasonable photoconductive properties. The basic process steps are (Des 65, Scha 71)

(1) sensitization of the photoreceptor by corona charging to fields greater than 10^5 V/cm

(2) exposure and latent image formation

(3) image development by triboelectrically charged toner consisting of pigmented polymer particles typically 10 μ in diameter

(4) image transfer, usually by electrostatic means with the paper in contact with image and photoconductor surface

(5) image fixing by thermal fusing of the polymer based toner to the paper

(6) removal of the remaining toner from the photoreceptor by mechanical means, and

(7) erasing the residual electrostatic image by uniform light exposure.

From an electrical standpoint, the following factors describe an idealized electrophotographic photoreceptor (Wart 69):

(1) The surface charge density required to reach a given voltage is proportional to the voltage, with the geometric capacitance per unit area as the constant of proportionality.

(2) Each photon absorbed results in the transport of one electronic charge unit completely through the photoreceptor.

(3) There are no mobile carriers that are not photogenerated. Thus there is no dark decay and no surface conductivity to neutralize the electrostatic image.

(4) The interface to the back-conducting electrode is a perfect blocking contact and does not allow charge injection into the photoreceptor or the accumulation of an interfacial potential drop.

(5) The top surface also is blocking and does not permit charge placed on the top surface to be injected into the bulk or to move on the surface.

(6) There is no change in the photoreceptor characteristics with cycling.

(7) The properties of the photoreceptor are the same everywhere on the photoreceptor surface.

No real photoreceptor meets this ideal and extensive discussion of each point, illustrated by Se, As_2Se_3 and As_2S_3, can be found in the literature (Tab 71). Nonelectrical requirements for a photoreceptor in this process are also demanding, and include uniformity and freedom from defects over large (5,000 cm^2) areas, low cost, and long operating life in a hostile environment.

Se, while not outstanding in meeting these requirements, represents a reasonable compromise. Inorganic amorphous semiconductors are not unique for this application. Crystalline photoconductors dispersed in a glass or polymer are widely studied. For example, ZnO, in paper, forms the basis of Electrofax process (Ami 65). Organic photoconductors have also received much recent attention: a poly-N-vinylcarbazole-2,4,7-trinitro-9-fluorenone charge transfer complex has been developed by IBM (Scha 71).

Optically Induced Phase Transitions

Aside from electrophotography, light can also be used to induce phase transitions in thin films of amorphous semiconductors, as has already been discussed in Section II on Structure and Bonding. This effect has potential utility in both the handling of images and in computer memories. The applications to image handling arise from the many ways in which the crystallized material differs from the amorphous, and awareness of these possibilities, which are mentioned in the Introduction, seems to stem largely from recent talks and discussions by S. R. Ovshinsky at ECD (Energy Conversion Devices, Inc.). At this time only preliminary aspects of that work have been publicly discussed. The modified properties include electrical conductivity and, therefore, the adherence of electrostatically charged toner. The modifiable properties, however, also include wettability, etch rates, the adherence of the illuminated material to adhesive tape, and optical properties. Thus the optically induced modification of reflectivity and absorption permits a form of photography. Other possibilities inherent in these properties relate to printing, photolithography, photoresists, and the direct generation of conductive paths under illumination. The material, which has been crystallized under exposure to light, can, if desired, be returned to the amorphous state through a heating and quenching cycle. This thermal erasure process can, in fact, also be accomplished through the use of light, in the form of a laser beam (Fei 71a). In the proposed application to digital memory, this erasure process is crucial. In the memory application it is also the same laser beam, at lower intensities, that would be used to probe the difference in optical properties.

As mentioned in Section II, it has been shown that incident light accelerates the growth of surface crystallites in amorphous selenium. Through the use of simultaneous electric fields, as well as through analyses plotting, the growth rates both as a function of absorbed light intensity and as a function of hole–electron pairs, Dresner and Stringfellow (Dr 68) make a convincing case for the fact that it is the hole–

electron pairs that account for the accelerated growth rate. They suggest that the hole, viewed as a broken bond, gives the structure the freedom required for crystal growth. It is not entirely clear whether the hole is only needed catalytically to reduce the activation energy for local reconfigurations or whether it is "used up" in the process. If it were, in fact, possible to use the same hole repeatedly in a catalytic fashion, we would have a more sensitive process with its own built-in gain mechanism. It would seem important to acquire an understanding of the mechanism as a guide in the search for more sensitive systems.

As a directly usable mechanism, the effect observed by Dresner and Stringfellow suffers from the fact that the enhancement in growth rate is proportional to the flux at low light intensities and becomes less sensitive at higher intensities. This simultaneous absence of a development step and a "threshold" means that the sample continues to be sensitive to stray light after its intentional exposure. One could, of course, use this mechanism in connection with a gating action, which elevates the sample to a higher temperature during the intentional exposure. Another way of obtaining a more sensitive mechanism with built-in gain would be to control the nucleation process, rather than the growth process, via light. Ovshinsky and Klose (Ov 71b) have, in fact, shown that the number of nuclei can be greatly increased by light, especially in films sensitized by the incorporation of certain catalysts. The nuclei can then be developed through subsequent grain growth induced either purely thermally or through a combination of elevated temperature and light. The grain growth can, if desired, be enhanced by the same light that has controlled the nucleus formation. Ovshinsky (Ov 71b) reports materials sensitive to exposures as low as 10^{-5} joules/cm^2. These films are self-fixing, i.e., grain growth at room temperature is negligible.

Another recent study (Bra 70) demonstrates an effect probably related to light-induced crystallization, accompanied by phase separation, in As–S and related glasses. These authors show that phase holograms can be recorded in such glasses. Exposure times with 10 milliwatts/cm^2 were of the order of minutes and the effect is probably too insensitive to be of immediate practical appeal. Refractive index changes continue for minutes after the light is turned off. This system also dispenses with a separate development process. As pointed out above, this is not an unmitigated advantage, since the system remains sensitive to light after its intentional exposure.

A reversible laser-induced crystallization in $Te_{.81}Ge_{.15}Sb_{.2}S_{.2}$ was observed by Feinleib et al. (Fei 71a). In this experiment, 3-μ spots were crystallized in times varying from 1 to 16 μsec, and with peak laser powers of 100 to 10 mW into the 3-μ spot. Two micron spots have also been observed in Se–Te films (Ov 71a). This work is at an early stage and considerable improvement may be expected.

The return to the amorphous state accomplished with the same laser (but at different power levels) (Ad 71a) is presumably a straightforward melt and quench process. Feinleib *et al.* (Fei 71a, Ad 71a), however, argue that the crystallization process is not a simple thermal process, since it is much faster than any other observed thermal crystallization process in these materials and occurs in a time comparable to that which the molten material spends in the same temperature range, while being *quenched.* In fact, Feinleib *et al.* go on to propose that even the electrically induced crystallization process is not purely thermal and is assisted by the excess electron–hole population. The area crystallized by the laser beam darkens visibly. Changes in optical properties can arise from the fact that the reflectance of the crystalline state below 4 eV in this material is approximately 50 percent larger than in the amorphous state (Ad 71a). The light scattering can also be changed by textural alterations in the film, which can include bubble formation (Fei 71b, Ov 71b).

The numbers discussed by Feinleib *et al.* appear to involve several photons per bond and thus no built-in gain mechanism. There is, however, appreciably more sensitivity than in the process discussed above in connection with hologram recording (Bra 70). The Se–Te films presently require 0.2 joule/cm^2 exposure (Ad 71a). This is somewhat higher than the range of sensitivity for diazo materials (10 to 100 millijoules/cm^2), which represent close to unit quantum efficiency. Diazo materials are probably at the lower edge of useful sensitivity for contact exposure systems, which, in turn, is a relatively undemanding application. The films described by Ovshinsky and Klose, which require a development step, require 10 μjoules/cm^2. The electrophotographic process requires about 1 μjoule/cm^2. The small size of the recorded spots observed by Feinleib *et al.,* combined with the microsecond time required for illumination, make this a contender for a beam addressed memory. This will be further discussed in Section VII on Technological Setting.

The various investigations reported above seem to have barely started to explore a rich variety of conditions for photocrystallization. These can involve variation of temperature during exposure, chemical sensitization, thermal pretreatment and development, the application of electric fields during illumination, and the role of the substrate.

Many properties of many materials are strongly affected by phase changes. The relatively small energy gap of the amorphous *semicon-*

ductors, which on the one hand permits photoexcitation of carriers, and on the other hand lowers the temperature required for reasonable crystallization rates, provides them with a unique range of possibilities in this image handling area.

ELECTRICAL DEVICES

The amorphous chalcogenide current-controlled negative resistance is sometimes called "Ovonic Threshold Switch" (OTS), the name used by ECD. The threshold switch and the bistable memory device are closely related, and both represent phenomena, which, with some variations, are in fact very widespread, although their physical mechanisms may be diverse. A review article oriented toward memory devices (Mat 71) gives a table with 23 different systems showing switchable resistances. Some of these entries are whole groups of chemically related systems. Not all of the settable resistances (memory devices) in this table are guaranteed to be reversible. In fact, it is possible in these systems to find a misleading first appearance of reversibility, which comes from a process in which highly conducting filaments are formed and then irreversibly burned out, leading to a quick consumption of the device material as new conducting filaments are formed.

The systems in Matick's table offer great diversity and include GaAs films, single-crystal yttrium iron–garnet, as well as semiconductor heterojunctions. It is not hard to find many additional possible entries, e.g., magnetite (Fre 69), Schottky barriers in GaAs (Es 70), and liquid Se and Te (Bus 70).

As already pointed out, all these systems do not necessarily utilize the same switching mechanisms. The behavior with respect to time constants, voltage and current levels, and polarity effects is sufficiently diverse to suggest that there are several mechanisms involved. Nor are these systems all equally likely candidates for the same set of applications. The diversity of system does exist, however, and is noteworthy in itself. Recent history has provided a heavy emphasis on the chalcogenides. Such a concentration on one aspect of a field is undoubtedly worthwhile because it extends and deepens physical insight. It would be unfortunate, however, to equate resistive memory devices and amorphous chalcogenides extensively and to let a judgment on one become a judgment on the other. Switchable resistances need not necessarily be chalcogenides, and chalcogenides have application potential that transcends electrically activated devices.

Threshold Switch Behavior

The typical behavior of this device, as shown in Figure 6a, is best summarized by a slightly condensed quotation (Sha 70b):

a. The I-V curve is symmetrical with respect to the reversal of the applied voltage and current.

b. The same switching characteristic is observed when the active material is sputtered, evaporated, or hot pressed between the [metallic] electrodes. It remains symmetrical even when the electrodes are of different contacting areas or of different materials.

c. In the highly resistive state, the conduction is ohmic at fields below 10^1 V/cm. At higher fields, the dynamic resistance, R_{dyn} decreases monotonically with increasing voltage: Typical values are 50 MΩ at 0.1 volt and 1 MΩ just prior to switching.

d. When the applied voltage exceeds a threshold voltage, V_t, the OTS switches along the load line to the conducting state. The transition time, t_t, of this switching process has been measured to be about 150 picoseconds. The shunt capacitance of a few picofarads and a lead inductance of about 10 nanohenries make it likely that measurements of t_t are limited by the package time constant. . . . The threshold voltage, V_t, is nearly proportional to the film thickness and can be adjusted by utilizing different materials to lie between two volts and 300 volts.

e. In the conducting state, the current can be increased or decreased without significantly affecting the voltage drop, termed the conducting voltage, V_c, across the device. Here the dynamic resistance is of the order of 10 ohms, an appreciable fraction of which is due to the lead resistance of the electrodes. V_c is nearly independent of thickness. By altering the electrode material, conducting voltage values between 0.5 and 1.5 volts can be obtained.

f. As the current is reduced below holding current, I_h, the OTS switches back to the original highly resistive state along the load line. The value of the holding current is typically between 0.1 mA and 0.5 mA, depending upon the composition of the active material.

. . . The OTS does not switch to the conducting state immediately as the threshold voltage is exceeded, but remains in the high resistance state for a period of time, t_d, called the delay time. The magnitude of this delay is strongly dependent on the amount by which the threshold voltage has been exceeded. t_d can be several micro-seconds for an applied voltage pulse of amplitude slightly greater than V_t or tens of nanoseconds for a voltage twice V_t.

. . . If current is rapidly removed from an OTS in the conducting state, a finite time is required for a device to recover its normal high resistance state characteristics. During this recovery time, t_r, V_t increases approximately exponentially. The characteristic time constant, τ, for the device to recover approximately 63 percent of its nominal threshold . . . is 3 microseconds for 30 volt units, 0.3 of a microsecond for 10 volt units, and should be considerably less for lower threshold voltage devices.

Fifteen volts are typical threshold voltages.

Current-controlled negative resistance (CCNR) has been seen in many systems, going beyond our previous list of internally relatively unstructured devices, and includes avalanche processes in gas discharges, e.g., neon bulbs. CCNR also occurs in many semiconductor structures, including p-i-n diodes (We 64), four-layer diodes, trigger diodes (Ti 69), and Reeves-Cooke diodes (Ree 55), in unijunction transistors, and in two terminal devices synthesized out of transistors (Ti 69). It is therefore not surprising to have a wealth of possible models, including the whole variety of mechanisms which have been invoked to explain nondestructive breakdown in insulators (Fra 56, Od 64). The situation in connection with the threshold switches is well characterized by Adler (Ad 71b): ". . . a large controversy still remains as to the mechanism for threshold switching in thin films of chalcogenide glasses." New mechanisms are being added as fast as old ones get demolished. A typical recent addition (Hom 71) invokes a filamentary plasma with high conductivity that travels from one electrode to the other. The present situation is too fluid to warrant the display of a complete set of rejected explanations. Many of the arguments and counterarguments involved can be contested by invoking geometrical nonuniformities. A plausible overview of some of the mechanisms is given by Adler (Ad 70, Ad 71b) and we shall here concentrate on the two most widely accepted sorts of explanation.

One of these, the "purely thermal" theory, pictures the rapid increase of conductivity with current that is implied by CCNR as a consequence of the strong temperature dependence of the conductivity found in these materials. The temperature changes in turn are a result of the heat dissipation resulting from the current flow. In this sort of explanation, one then invokes only macroscopic quantities: thermal conductivities, heat capacities, and the temperature dependence of the electrical conductivities.

The other kind of theory invokes microscopic concepts including the injection of charges, the redistribution of charges in traps, the modulation of barriers at the surface of the amorphous semiconductor, and the modulation of the effective mobility and of the internal field distribution through trap filling. Plausible recent versions of such theories stem from Fritzsche and Ovshinsky (Fr 70a), and from Henisch, Fagen, and Ovshinsky (Heni 70). Earlier versions stem from Mott (Mo 68, Mo 69b, Mo 69c). These are qualitative pictures. The detailed elements, which enter into such a theory, e.g., the existence or nonexistence of barriers at the electrodes, are not available with any confidence from other investigations. Van Roosbroeck (VanR 71) has pointed out that injection effects in a material with a dielectric relaxation time larger than the recombination time are very different from those typically seen in crystalline semiconductors and has provided a theory for both switching and storage devices on this basis.

It is becoming relatively widely accepted that for thicker samples and higher temperatures the purely thermal picture applies, while for the thinner samples and lower temperatures an electronic injection mechanism must also be involved. At room temperature, films thicker than $10 \ \mu$ are generally believed to exhibit purely thermal breakdown. Unfortunately, it is not clear that the thermal theories that are most often invoked really apply.

Lueder and Spenke (Lue 35, Spe 36) have given a theory of thermal breakdown. Their paper contains information that does not appear to have been fully appreciated by all subsequent authors. In particular, Lueder and Spenke show:

(1) For the typical capacitor geometry with thermal gradients and current flow parallel, and with the electrodes held at constant temperature, there is *no* CCNR but only a limiting voltage which is approached asymptotically at high currents.

(2) For this same geometry, a CCNR results if the electrodes are not kept at constant temperature but connected to the heat sink through a thermal resistance. One would thus expect to see expressions for threshold field which depend on this additional thermal impedance.

(3) If the capacitor geometry device, assumed to be conducting uniformly over its area, shows a CCNR, then there is a spatial instability leading to filamentary current flow, and this can drastically sharpen the drop in voltage with increasing current seen in the negative resistance region. The point of onset of the spatial instability depends on the relative ease with which heat is transported from one part of the device to another, as compared to the transport of heat out of the device altogether.

These features are supported, in part, by the work of others (Böe 70b, 71, Key 71, Od 64).

There seems to be a need for a more adequate thermal theory based on modern device geometries, which (Lue 35, Spe 36) takes into account the exact point at which spatial instabilities commence, as well as the strong dependence of conductivity on electric field observed in these materials. The work of Volkov and Kogan (Vo 67) concerns itself with the formation of lateral instabilities but does not seem to have been generally understood. There is room for a simpler theory closer in spirit to the existing thermal theories. Recent calculations by Warren and Male (Warr 70), and Böer (Böe 71) deal with a number of the above points adequately and report better agreement with experiment for threshold field and switching delay time respectively. Other thermal calculations are currently reported to be in progress and it can be hoped that the existing gap in theory will soon be closed.

The experimental data that bear on the thermal model vs. the space charge modulation models are primarily of three kinds:

(a) Thickness dependence of threshold field (Ko 69). The threshold field is independent of thickness for thin samples and decreases with thickness for thick samples. This thickness dependence is invoked (Ko 69, Fr 70b, Sto 70) to show agreement with thermal theories for thick samples ($>10 \mu$). Since a really good thermal theory is not used in these comparisons, they provide suggestive, but incomplete, evidence. The improved theory of Warren and Male (Warr 70) claims to explain the threshold fields in films as thin as 2 μ.

(b) An effect due to the polarity of the preceding pulse on the threshold field would be expected only for an injection model, not for a thermal model. Shanks (Sha 70a) and Balberg (Bal 70) find no such polarity effects, whereas Henisch and Pryor (Heni 71) find such effects, though modest at room temperature. These are nonexistent at an above-room temperature in some devices. At $-78°C$, the polarity effects become appreciable (Pr 71). Although more evidence is needed, there seems to be at least some indication for the presence of charge injection effects.

(c) Measurement of total injected charge required for switching. Haberland (Hab 70) considers the delay in switching as a function of applied field and thus provides some evidence for the need of a minimum accumulated charge as a condition for switching. He discriminates against a mechanism requiring elevation to a critical temperature by showing that more energy input is required for switching at high temperatures than at low temperatures. As in (b) above, the evidence is not as clear as one would like but favors a model in which injection plays some role. In contrast to Haberland, Csillag and Jäger (Cs 70) report a switching process in which the accumulated energy seems to be the critical parameter.

The Memory Device

Typical memory device characteristics are shown in Figure 6b. The basic action of the memory device is best described by a slightly condensed quotation from Helbers (Hel 71):

. . . The normal amorphous . . . phase is an insulator, having a room temperature resistivity of 5×10^4 ohm-cm. If, however, the material is heated above a critical temperature and cooled slowly, a second micro-crystalline phase will form. This material behaves as a degeneratively doped semiconductor having a resistivity of 0.3 ohm-cm. The material may be easily returned to the high resistance state by once again heating it beyond its glass transition temperature and cooling very rapidly, thus freezing the material in the disordered or amorphous phase. The threshold switching action provides a conducting path through the active material, which allows subsequent deposition of the required transformation energy in a confined region using *reasonable* currents and voltages . . .

. . . The SET pulse is a high-voltage pulse (typically 25 volts) which is in excess of the memory device's breakdown or threshold voltage. This pulse will cause the memory switch to fire and conduct current. This current is limited to a few milliamperes by a high pulse source impedance. The SET pulse is fairly long (typically 10 milliseconds) so that not only the active region of the switch is heated but also the immediate surroundings. When the current is stopped, the active region cools slowly; the active material is thus left in the micro-crystalline low resistance state. In order to make this slow cooling process more reliable and predictable, a trailing edge is normally added to the SET pulse. This trailing edge provides a gradually decreasing current density in the device and thus insures the slow cooling of the active region. It should be noted that if the SET pulse is terminated a few hundred microseconds after the onset of conduction, the memory switch will auto-matically revert to the OFF or blocking state. This is due to the fact that insufficient time will have elapsed for nucleation of the micro-crystalline phase of the active material.

To return the memory switch to the OFF or blocking state, the active ma-terial must be converted back to the amorphous or glassy phase. This is ac-complished by applying a short pulse of high current to the device, typically 150 milliamperes for 5 microseconds. This pulse will heat only the active region of the device causing a very steep temperature gradient. Thus, after the current pulse ends, this steep temperature gradient will cause a rapid cooling or quenching of the active region; freezing the material in the OFF state. The memory switch then, actually turns OFF *after* the RESET pulse; not during it. . . .

There is, in fact, much more widespread agreement on the above picture than there is on the mechanism for the threshold switching action which precedes the crystallization process. It has, however, not been clearly proven that the crystallization process is entirely thermal. As already indicated in the earlier section on optical effects, it may be supplemented by the effects of nonequilibrium hole–electron populations. There have also been observations (Ver 57) of direct electrical effects on crystallization in anodic tantalum oxide.

The above description may be supplemented by additional char-acteristics. The crystallization process is accompanied by phase separa-tion, as discussed in the section on structure and bonding, and the filament consists of Te-rich needle-shaped crystallites (200–400 Å) (Si 71). The glasses that yield memory devices rather than threshold switches are materials which have lower glass temperature, i.e., they crystallize more readily. As in almost any memory device, there is likely to be some compromise between ease of switching and the stability of the terminal states. The crystallized filament has a diameter of 2 to 3 microns (Si 71). The current density in this thin filament during the reset operation far exceeds that which can cause electromigration (Ame 70) in *room temperature* metals. As pointed out by Pearson (Pea 70), the crystallization (as well as the quenching) is not necessarily homogeneous throughout the filament, but could be dependent upon making and destroying bridges between regions that remain highly conducting. Even if the filament is a uniformly crystallized region, its diameter can vary, depending upon the exact previous electrical history (Hel 71). We thus do not necessarily have a memory device with two well-defined, self-stabilizing states between which one can cycle back and forth. (In reliability tests of these devices, it is important therefore to have a diversity of test patterns, and not to depend on just one very simple highly periodic sequence.) The range of excitation permitting stable operation has been investigated (Heni 71, Nea 70). In the reset process, a region about twice as wide as in the set process becomes liquid and is cooled too quickly to permit crystallization in the surrounding region (Sie 71).

For the further development of the memory device, the microscopic nature of the filament, its changes during switching, and the effect of the filament structure on the electrical characteristics need to be studied in more detail. The presence or absence of electromigration and its effect on characteristics must also be determined. The role of electromigration as a failure mechanism is currently under study at ECD, together with some of the other kinetics and geometry of phase transformations in the filament.

Radiation Hardness

It is clear from the literature and from comments made to the Committee that amorphous chalcogenide negative resistance devices can have a very high tolerance to radiation exposure. The question of radiation hardness in systems is an exceedingly complex one. It involves classified material specifying a meaningful degree of exposure to various kinds of radiation. The radiation immunity question, furthermore, cannot sensibly be addressed at the component level but stresses systems questions. We shall, therefore, content ourselves with pointing out that there are some positive data. We must also stress the distinction between memory devices (on which less data have so far been published) and threshold devices. Threshold devices, combined with passive devices, can perform all digital logic functions and perhaps most other required electrical circuit functions. Such circuitry, based entirely on two terminal devices, is, however, often awkward and is delicate in terms of its sensitivity to device parameters. Radiation hardness questions can be separated into questions concerning permanent structural change in the device involved, and questions concerning the reliability of operation during radiation exposure due, for example, to radiation-induced hole–electron pairs.

The available literature is cited in the reference list as (Ov 68b, Eva 68, Henc 70, Fl 70, Sha 70b, 70c, Smi 71, and Ni 71). The last three of these discussions are the only ones with bearing on memory devices. Nicolades (Ni 71) asserts that 40 percent of the memory samples initially in the "on" state converted to the "off" state during irradiation at an integrated neutron flux of 10^{15} neutrons/cm^2. Smith (Smi 71), however, finds that ECD's memory devices are relatively insensitive to an exposure in excess of 10^{16} neutrons/cm^2. This reference list does not do justice to the level of insight into the radiation hardening question that exists at Picatinny Arsenal and the Naval Ordnance Laboratory, but which at this time, at least, has not reached a convenient publicly available form.

The above section should not be construed as an indication that amorphous chalcogenide devices are, in a very general sort of way, insensitive to other aspects of their environment. For example, the threshold voltage of memory devices can change from 25 volts at 20°C to 4 volts at 70°C.

VII

Technological Setting

SCOPE

The preceding section covered device physics. In this section we shall compare the resulting devices to some of their competitors. Our purpose here is not that of providing guidance to the technologist, much less to the device user, but rather to sketch the setting in which this new field finds itself. While new technologies can create their own special opportunities, it is instructive to view them in the context of competing technologies. It is in this spirit that we will provide some comparisons with other technologies.

The electrophotographic applications are, on the one hand, so well established, and, on the other hand, their future potential so intimately coupled to systems questions and to proprietary information that further discussion here would be difficult. The image handling potential through selective photocrystallization is at such an early stage that again very little can be said. We shall, therefore, restrict this discussion to the threshold switch, the electrical memory device, and the optically accessed memory device.

75

THRESHOLD SWITCH

As indicated before, the threshold switch is one of many negative-resistance two-terminal devices. Three-terminal devices are generally preferred to two-terminal devices, with the same general capabilities for control and switching applications, because they provide a more natural isolation between control signal path and the controlled signal, and seem to be less demanding in terms of the required device parameter tolerances. Special performance capabilities or ease of fabrication may, nevertheless, make two-terminal devices preferable in some applications.

The most widely discussed application for the threshold switch (Ad 70, Heni 69) is in a display panel. In this case, the threshold switch is used to control the excitation supplied to an a.c. electroluminescent light emitting element. The electroluminescent element, acting largely as a capacitance in series with the threshold switch, becomes a bistable circuit in its mode of response to an impressed a.c. voltage. It can be set in an "on" or "off" mode and left there until reset. The display, therefore, has a built-in memory and need not be continually regenerated from an external buffer memory as long as the image content is left unchanged. The use of a.c. electroluminescence in displays is, of course, not new (Gre 64) and has been tried with a variety of control mechanisms. It is the use of the threshold switch for this purpose that is novel.

This display scheme is in competition with many other old and new proposals, some oriented principally at the large area display, i.e., at replacing the cathode ray tube (CRT), and others oriented more at the display of a relatively modest number of alphabetic and numerical characters. The large area display proposals frequently still involve an electron beam but go beyond the cathode ray tube to provide some kind of storage mechanism built into the display itself. These proposals include the photochromic storage CRT (Me 70), the cathodochromic storage tube, secondary emission charge storage tubes, and others, many of which are reviewed by Kazan and Knoll (Kaz 68). Other recent display proposals include the use of liquid crystals, semiconductor light-emitting devices, ceramic ferroelectrics, and cylindrical magnetic domains (Al 71). Uniformity of light output over the area of a display is easily achieved in an undeteriorated CRT, where the phosphor on the CRT face has come from one uniform batch. Uniformity is not always equally easily achievable in proposals involving a discrete device for each picture element and where light output can be a very highly nonlinear function of electrical excitation.

Numerical displays have long been commercially available in the form of gas discharge tubes with shaped electrodes. More recently miniaturized incandescent filaments have been developed for numerical displays. Multidigit gas discharge panels, in which the information is fed in at one end and then shifted along (analogous to the famous news display at Times Square), have also appeared (Har 70). Progress in a number of display technologies has been reported at a recent conference (SID 71). The technology that perhaps most closely resembles the threshold switch combined with an a.c. electroluminescent phosphor is the plasma display technique of Bitzer and Slotow (Joh 70, SID 71), which also invokes an array of bistable elements, excited by a.c., each consisting of a CCNR (current controlled negative resistance) in series with a capacitor. In the plasma display, however, the CCNR is that of the light-emitting gas discharge and the capacitance is that of the passive walls of the element. In the current version of the plasma display, where no walls separate adjacent light-emitting elements, uniformity over the display area is readily achievable.

Many of the more recent display proposals (including the one utilizing the threshold switch) are characterized by (a) the many peripheral connections characteristic of a rectangular matrix array, rather than the few terminals of a scanning device such as the CRT tube, and (b) built-in memory. The pros and cons of these alternatives are complex and beset with overall systems considerations.

The plasma display panels require somewhat higher voltages than the electroluminescent devices. On the other hand, a.c. electroluminescent devices suffer from a well-known deterioration of light output during operation (Iv 63, Sy 68), which has thus far limited their really widespread utilization. Energy Conversion Devices, Inc., in a contract report (Van 70), reports progress toward obtaining longer-life phosphors. If this result is obtained under reasonable operating conditions with good light output, it is of significance quite aside from questions regarding the utility of the threshold switch.

THE ELECTRICALLY ADDRESSED MEMORY DEVICE

Information can be held in continuous media (often called storage) or else in discrete devices (often called memory). Magnetic tapes, disks, and proposals for reading and writing with optical beams are all examples of continuous media. The discrete category is typified by magnetic cores and integrated circuit memories. It is this latter group into which the amorphous chalcogenide memory (hereafter called ACM) device fits, and which we will consider first.

Attempts to make integrated random access memories out of magnetic elements has led to discouragement (Pet 70), and most current proposals for random access memory developments center on the utilization of a silicon chip. The devices that can go on the silicon chip, however, show continually increasing diversity, and now also include the chalcogenide devices. In our discussion here, *we will by no means try to survey the full flexibility of the silicon technology* but only allude to some typical schemes. Deposition on silicon permits the realization on the same chip of peripheral address decoding circuitry, which selects the particular memory lines being addressed, and thus reduces the number of interconnections to the chip. Deposition on top of silicon also permits auxiliary devices, required for an information holding cell, to be made on the chip. Thus, for example, ACM devices require a series diode with each memory device. This is needed so that a high-resistance device will not be unintentionally shorted out by a group of neighboring low-resistance devices. If there are series diodes present, then each potential "shorting" path will have to traverse at least one diode in its reverse direction.

The chalcogenide devices, as has already been stressed, presently require about 10 milliseconds for the transition to the low-resistance state. Since the device reacts somewhat unpredictably to switching pulses, a "verify" interrogation after each switching attempt is recommended (Hel 71); if switching has not taken place, then another switching attempt is made. These features have led to a designation of ACM as a "Read Mostly Memory," i.e., as one that has to have its information content reset only very intermittently. Such memories are certainly useful, particularly as control memories which help set a computer's internal information flow structure. They can also be used, for example, in encoding arrays. On the other hand, it must be remembered that a special technology is not needed for this purpose: an ordinary read–write memory does not necessarily have to be written very often. Thus we will, in our comparisons, compare amorphous chalcogenide devices to a number of ordinary memory candidates, as well as some more specialized proposals.

Amorphous chalcogenide devices do have two possible advantages over most of the comparison candidates. Their radiation immunity is one. The other is that ACM's hold information in a more permanent and less volatile form than most of the other comparable devices. ACM does not require power to maintain information; there is no indication that stored information degrades during storage, and it can be interrogated nondestructively, i.e., without an intentional switching process. This lack of volatility is certainly desirable, all other things being equal. On the other hand, lack of volatility is generally crucial only in the really large information arrays kept on disks, drums, and tapes. Computer memory proposals have, in recent years, pushed in the opposite direction: they have gone to *more* volatile schemes for the sake of gains in cost and density. There are, however, applications, particularly in military systems, where a premium is put on nonvolatility and nondestructive readout.

As a first comparison candidate, we shall refer to plated wire memories that store information in the magnetization of a film (McC 70, Ms 70, En 70). These magnetic memories can be read in access times comparable to the amorphous chalcogenide memories (70 ns), are read nondestructively, and are nonvolatile. Write cycles for plated wire memories are also in this same nanosecond range rather than in the millisecond range, which is characteristic of the ACM. The question of relative costs is open, but the ACM does have the advantage of higher density and potential integrability with peripheral circuits on the silicon chip, which would not be easy for the magnetic arrays. Furthermore, the ACM requires somewhat less drive current in its switching process. On the other hand, the plated wire memories with their fast write capability have broader application potential, and this can have a strong effect on cost.

Our remaining comparison candidates are all silicon chip technologies. Within these technologies, density (bits/in^2) is the key to cost. Processing costs for silicon wafers may vary by modest factors, depending on device details. Densities, however, can change over several powers of ten, and density is, therefore, the most important single parameter. Currently, ACM memories (Nea 70a, Hel 71) can be made with 256 bits, without peripheral circuitry on a chip 116 mils $\times$ 126 mils, or 17,500 bits/in^2.

All of the densities to be cited, for both ACM and the other technologies, are subject to improvement. For example, current ACM's as available from ECD, and developed by ECD and Intel, have the chalcogenide device and its series p-n junction side by side. A more refined version places them on top of each other. Nevertheless, there is a basic distinction between ACM and the other silicon technologies. For all the other approaches, current flow and power dissipation in the device can be reduced as the device cross-section is reduced. The ACM device, however, in its low-resistance state uses a filament cross-section independent of the device area, and neither its heat dissipation nor its current demand can be decreased by analogous techniques. It is, therefore, important to discover ways of reducing ACM current requirements. Since the present chalcogenide device involves very high reset currents, high-device densities will probably require such a current reduction. Even at the present density (Nea 70a, Hel 71), steps had to be taken to prevent an excessive voltage drop within the diffused semiconducting regions, which are used to give access to the devices in one of two perpendicular directions leading into the array. These semiconducting "lines" could, of course, be replaced by much more highly conducting metallic lines, resulting in a chip with two layers of perpendicular transmission lines insulated at their cross-overs. This is a more complex and expensive technology but certainly one worth considering. But even with the metal lines, with densities of 10^6 bits/in^2 (potentially feasible in some of the other technologies to be discussed), there would be difficulties (Mat 71). This density would require the feeding of 200 mA through a line perhaps 0.2 mil by 1 μ thick. The high current density of 4×10^6 amps/cm would then lead to electromigration problems in the metallic paths (Ame 70). Of course, electromigration need not be a problem if the memory is to have its content reset only very occasionally and thus is subject primarily to the much smaller read currents.

In comparing bit densities for the various silicon chip technologies, it must be borne in mind that most current memory developments work with chips in the same general size range (somewhat over 0.01 in^2). As a first approximation, it is therefore reasonable to compare bits/chip instead of bits/in^2. A number of recent collections of papers on integrated circuit memories (ss 70, IE 71, Com 71, ssc 71) yield the following chip densities, to be compared to the current 256 bits/chip for ACM (without decoding on the chip in the case of ACM):

(a) 128 for a high-performance bipolar transistor memory cell, with decoding on the chip. [One company (Cog 71) has announced a 512-bit chip, and a trade press publication refers to another company's plans for a 1,024-bit chip (El 71a).]

(b) 256 for a high-performance bipolar transistor memory cell, without decoding on the chip.

(c) 256 for Insulated Gate Field Effect Transistor memory cells, with decoding on the chip, and using six devices in each cell.

(d) 1,024 for an Insulated Gate Field Effect Transistor circuit, using 3 devices per bit, and including peripheral circuitry on the chip. This circuit does not have steady-state bistability, and its content must be refreshed every few milliseconds. This is typical of the extent to which an extreme degree of "volatility" is accepted for the sake of density. [A trade press column (El 71b) refers to a 1 device/bit cell, leading to a 2,048-bit chip.]

(e) 8,192 bits for Insulated Gate Field Effect Transistors used in a Read *Only* Memory. Unlike the ACM, this device has its content permanently set in the factory. [A trade press report (El 71c) refers to 8,192 bits also obtainable with bipolar transistors.]

All of the above densities (a)–(e) can be expected to increase as the state of the art of photolithography used in the manufacturing process improves. Densities closely approaching that of (e) above can be reached in other ways without sacrificing the ability to change the information content of the memory. In several recent proposals (ssc 71, San 70, Eng 70), charges are moved across a silicon surface, under the control of electrodes applied to an oxide over the silicon, and without taking the current in and out of the silicon. These devices thus store information (presence or absence of a charge) in a delay line, without taking the charge in and out of the silicon surface. One to two square mils per stored bit are required. These charge transfer proposals do not permit random access, since the bits are passed sequentially through the read apparatus. Thus very long delays are involved in the readout process. These devices also lack steady-state bistability, and information must be continually circulated and regenerated to prevent it from disappearing. Furthermore, the intrinsic device speed (transfer of one bit) is also somewhat lower than in the transistor arrays listed in (a) through (e) above. Very similar systems compromises are involved in the use of cylindrical magnetic domains or "bubbles" (Bon 70). These are also used in a delay line scheme, and also involve basic time constants long compared to those of the transistor memories. Bubble memories have been operated at data rates of 10^6 bits/sec. (Fisc 71), and with bit densities in the $10^6/in^2$ range (Shi 71). Bubble memories are volatile only if the d.c. bias field they require is supplied through an electromagnet.

We shall now go on to describe three devices probably utilizable on a silicon chip that have the potential of yielding very high densities without sacrificing random access, and which require at most a very limited sort of regeneration. All three devices are relatively novel and are, perhaps, less clearly understood and under control than the chalcogenide devices. All of these devices have the potential of allowing bits to be about a mil apart, i.e., giving densities $\sim 10^6/in^2$ with a silicon technology whose smallest controllable surface dimensions are 0.2 mil. As the manufacturing capability in silicon progresses, all of these technologies can be expected to advance in density.

The Metal–Nitride–Oxide–Semiconductor Device

The Metal–Nitride–Oxide–Semiconductor Device (MNOS) (Fro 70) is a field effect transistor in which the gate insulator is a two-layer sandwich. A thin SiO_2 layer next to the silicon has a thick Si_3N_4 layer superimposed. The gate electrode can cause tunneling back and forth between states at or near the interface of the two insulators and the silicon surface, and thus can "set" the device in one of two modes of operation. This device, in the reading process, can be about as fast as the other high-speed silicon technologies we have discussed. It is not equally clearly a fast "write" device, requiring a fair fraction of a microsecond in a system with reasonable parameters. Like ACM, it involves voltages (20 to 40 volts) that are higher than those required by transistor memories and is, therefore, more demanding on the peripheral circuitry. But, as in the case of the ACM, these voltages are certainly within the range of achievability. The charges at the interface can leak away slowly over a period of perhaps months. Thus the memory is not completely nonvolatile. Only one device per memory bit is required. The area per bit has not been clearly defined as a result of a detailed proven design, but would not be far from the optimum obtainable for a given silicon capability, e.g., 1.0 square mil for a 0.2 mil resolution capability (Mat 71, Te 70).

Heterojunctions

Memory effects have been observed in ZnSe–Ge heterojunctions, in which the resistance can be switched from low to high values (Hov 70, 71). Similar, but less well-studied effects have been observed in GaP–Si heterojunctions. Of these, the GaP–Si system would, of course, be of greater technological appeal since the integrated silicon technology is much more highly developed than the corresponding germanium technology. Switching times into the low- and high-resistance states are respectively 100 nsec and 10 nsec. Switching voltages and currents are of the order of 1 volt and 1 ma. These devices are, therefore, fast and low-powered compared to ACM. Unlike ACM, switching in the heterojunctions requires polarity reversals. This causes more demanding specifications on the diode in a memory, which will have to be put in series with each device. Uninterrupted switching of the heterojunction at frequencies above 500 kHz causes problems. The information state of the heterojunction is not as unconditionally stable as in the ACM and it may require some very occasional regeneration. The simultaneous appearance of strong polarity effects, switching ease, and poorer information stability suggests more purely electronic phenomena than in the chalcogenide device. The heterojunction technology, like the MNOS device, could be expected to lead to a near minimum cell area for a given state of silicon technology. The device does seem to depend on some kind of filamentary action and is, therefore, not completely "scalable." On the other hand, the filament current is much smaller than in ACM and, therefore, should be less of a problem. Some "forming" treatment is typically required by these devices, more clearly than by the chalcogenide devices, but not as clearly as for the next device to be discussed.

$Nb–Nb_2O_5–Bi$ Junctions

Switchable resistances have also been observed in anodized amorphous (i.e., crystallites <20 Å) Nb_2O_5 with a bismuth counterelectrode (Herr 71, Bas 71). A real "forming" process is involved resulting in a permanent filament of about 2 μ^2 cross-sectional area in which structural changes have occurred. Just as in the heterojunction, device polarity reversal is required. The switching time into the high-conductance state is 10 nsec or less. Switching time into the low-conductance state is a complex function of forming treatment and previous pulse history but probably can be kept below 1 μsec. Switching voltages are about 1 volt, as in the heterojunctions. The switching currents are still smaller, of the order of 0.1 ma. The stability of information in these devices has not yet been well characterized but there are indications that the information can deteriorate over a period of weeks. As in the case of the heterojunctions, all of this is suggestive of a more purely electronic mechanism than in ACM.

OPTICALLY ADDRESSED MEMORIES

Any consideration of optically addressed memories must take into account the prevailing technology: magnetic surface recording that is written and sensed through flux changes. (Some of the optical memories also invoke magnetization changes, but we shall use "magnetic recording" to signify this classical inductive form.) Magnetic recording utilizes tapes, drums, and disks, as well as a number of other geometrical forms. As in the case of the wired memories, bit density is a key index to the usefulness of the technology. Optical technologies are limited by the wavelength of light. The achievement of greater density in magnetic recording has been a slow and painful process. Development activity is directed toward obtaining smaller dimensions for the thickness of the magnetic layer, for the gap in the head, and for the spacing between head and magnetic layer, all without risking too much wear between head and surface. This involves ingenuity in signal coding methods, which eliminates the need to record through the whole thickness of the magnetic layer.

Where is this technology currently? An existing commercial disk system (PP 70) has about 4,000 bits per inch along the track, 200 adjacent tracks per inch, giving an areal density of $0.8 \times 10^6/\text{in}^2$. Data is read out from one head at about 6.4×10^6 bits/second. Advanced existing tape systems (Hu 70, Dam 68) have very similar areal densities and data rates.

It is widely accepted that progress in magnetic recording beyond this status is possible. One prediction (Hopn 69) anticipates densities of 5×10^7 bits/in². Densities along the track as high as 40,000 cycles/in were achieved in audio recording over a decade ago (Bro 60). In more modern digital applications, 16,000 bits per inch (Auf 70) have been achieved. The miniaturized heads required by greater track density may also lead toward new methods of head manufacture, more akin to semiconductor processing technology than to mechanical assembly (Rom 70). Such integrated head arrays could provide the capability of placing one head over each track, thereby eliminating the need for track-to-track motion. (This is routinely done today in *low* track density drums and disks.)

Optical storage proposals are either holographic or direct digital. The holographic schemes handle large digital arrays in parallel and this gives them a great deal of system appeal (Ra 70). The parallelism, however, also leads to a more demanding set of requirements than when handling only one bit of information at a time. At the current time, even the direct digital schemes have shown feasibility in only a marginal way. With this in mind, we shall, in the following discussion, orient ourselves more toward the direct digital schemes.

Direct digital optical schemes are investigated largely because they offer promise of a higher track density and also perhaps permit quicker switching and servoing of the transducer in its track-to-track motion. They may not be, in the long run, competitive with magnetic recording in density along the track. For a given area density, a high density along the track is more desirable since it gives higher data rates (for the same relative velocity between transducer and recording medium) and fewer tracks between which switching is required. Thus it is not enough for optically accessed memories to provide higher track densities at the expense of a lower density along the track. The optical scheme must provide either a much faster track-to-track switching time (e.g., by replacing mechanical motion with electro-optic or acousto-optic deflection at modest cost) or else must provide a *much higher* density of bits.

The general resolution capability of optical schemes is well illustrated by a storage system called Unicon that is being developed by Precision Instruments (PI 70). In this system, a laser is used to vaporize, or perhaps melt, spots in a metal film. This is a scheme in which information cannot be erased, once written. All other things being equal, one would, of course, much prefer a reversible storage medium. In this system, bits are about 3 μ apart along the tracks and 8 μ apart between tracks, giving about 25×10^6 bits/in². Three microns may well be the limit for a practical system in which a depth of focus has to be preserved in the presence of large moving surfaces. Furthermore, the 8 μ is also likely to be characteristic of the fact that for reasonably easy track servoing, the tracks should be further apart than the bits along the track. The data rate is 4×10^6 bits/sec. It is therefore the track density, and only the track density, that is impressive compared to current magnetic schemes.

As another standard of comparison for optical technologies, we may also mention a kind of ultra-high density phonograph record developed for home display of previously recorded television programs (Gil 70). It is one of several technologies currently being developed for this purpose (Boy 71). This disk is stamped in a factory. It is not recorded in real time or erasable. It is thus representative of a much more limited technology than the others discussed in this section. This disk, however, has an information density of 3×10^8 bits/in² (using an FM scheme with more than one bit of information per hill and valley).

A great many proposals for reversible optical media have been made utilizing photochromic effects, photoconductor–thermoplastic combinations, color center alignment, and optical damage in $LiNbO_3$. We shall here consider only a typical subset of relatively active contenders. These are Ferroelectric–Photoconducting (FE–PC) combination sandwiches, thermomagnetic schemes, and the amorphous chalcogenides.

The FE–PC schemes (Chapm 70, Ken 70a, Ken 70b) use light to control a photoconducting current, which, in turn, controls the switching of a ferroelectric. The ferroelectric can be read electrically (Chapm 70) or optically (Ken 70a, Ken 70b). It should be noted that 1.26-μ gratings have been recorded in the ferroelectric, demonstrating the high resolution capability of the scheme. This scheme can, in principle, be very sensitive in the writing process, since it can utilize the gain of a photoconductor. It is, therefore, a particularly likely candidate for either holographic schemes or schemes that dispense with the laser and its accompanying complex optical deflection methods in favor of the readily deflectable spot on a CRT. It is also a system that may suffer from the long decay time of the photoconductor. However, this need not be a crucial problem, with a sensible overall systems design. This proposal involves two technologies (ferroelectrics and photoconductors) that have both been beset by a long history of difficulties in precision applications, and it is likely to require a major materials effort.

The thermomagnetic schemes (Esc 70, Mag 70, Hu 70) use a laser beam simply to heat a material to or through a Curie point or compensation temperature at which an applied magnetic field can switch the material. Reading is done magneto-optically. A particular sensitive magneto-optic proposal has been developed utilizing doped EuO (Far 71, Pat 71), but at the expense of having to keep the medium at liquid nitrogen temperatures. The EuO memory has been demonstrated under conditions resembling those of a practical memory with densities along the track of 3,000 bits/in, areal densities of 4×10^6 bits/in^2, and data rates of 5 Mb/sec. It is probably the only one of the optical schemes that has been shown to operate at rates comparable to those already existing in commercially available magnetic recording. The EuO scheme can also be utilized with a GaAs laser that exists in the form of integrated arrays of many lasers. Thus the EuO scheme can eliminate or minimize the need for light deflection schemes or mechanical transducer motion.

The amorphous chalcogenide proposal (Fei 71, Ad 71a) has already been discussed. The fact that the laser written spots are very small (2 to 3 μ) is attractive. The time and energy requirements that have been demonstrated are not yet very impressive, but the technology is at such an early and poorly understood stage that one can hope for major improvements. Amorphous chalcogenide films can be given much greater sensitivity (Ov 71b) if a subsequent development step is allowed. While a development step is not easily included in high-speed digital equipment, it is not completely out of the question, especially if it can be accomplished in milliseconds.

In addition, the following points should be noted:

1. The change in refractive index, which is available for detection in the readout process, is much larger for the amorphous chalcogenide films than for the magnetic or ferroelectric devices.

2. In any complete or partial thermal scheme involving, for example, thermomagnetic or crystallization methods, in which thin films are continually and rapidly cycled over appreciable temperature excursions, there must be some concern about the effects of thermal strain on the adhesion and integrity of the thin film.

3. The amorphous chalcogenide crystallization process has so far resulted in far cleaner demonstration experiments than any of the other optically addressed materials (Ov 71a).

Order-of-magnitude estimates for the sensitivities during the writing process are as follows:

(a) Crystallization of amorphous chalcogenides 10^{-1} joules/cm^2
(b) Europium oxide 10^{-1} joules/cm^2
(c) FE–PC observed (with a photoconductor gain well below
    unity) 10^{-3} joules/cm^2
(d) Amorphous chalcogenides with development 10^{-5} joules/cm^2
(e) FE–PC calculated (on the basis of observed photoconductor and
    ferroelectric properties, but not in actual sandwiches) 10^{-6} joules/cm^2

In connection with these figures, it must be remembered that a purely thermal scheme, such as that involved in EuO storage, is not susceptible to really dramatic further improvements in sensitivity. The erasure process in (Fei 71) may also fall into the same category.

Sensitivity is a critical parameter only if the available light is limited. In an arrangement having only one laser per system, writing on only one spot at a time, sensitivity may not be a critical problem.

VIII

General Observations and Recommendations

GENERAL SCIENTIFIC INTEREST

The field of amorphous semiconductors is sufficiently challenging and difficult to occupy experimentalists and theorists for some time to come. It is important both in connection with the general study of disordered systems as well as the established usefulness of glasses. Indeed, there has been considerable growth of the R&D effort and a marked increase in the scientific interest in the area of disordered materials during the last few years. This comes appropriately at a time when, after forty years of concentrated effort, a remarkably high level of scientific understanding and technical use of crystalline materials has been achieved.

1. *In view of the absence of a fully developed conceptual framework for the science of amorphous semiconductors, the Committee believes that close collaboration between experimentalists and theorists is essential for further progress in the field.* Experimentalists should be encouraged to choose investigations that might differentiate between various theoretical models. Similarly, theorists should concentrate their efforts in areas where results are subject to experimental verification.

2. *Although the development of simplified models has been extremely useful, it is desirable that more attention be focused on calculations of*

electronic potentials, energy levels, and transport properties directly applicable to physically realized glass structures. Initially, these would require large-scale computer calculations. As pointed out in this report, such programs are now in their beginning stages. Hopefully, it will be possible to extract from this body of experience some simpler approaches and models that yield semiquantitative results which can be compared with experimental data. It is particularly important to inquire how the short-range order in an ideal glass determines the band structure and whether there are features of the density-of-states near the band edge that are general for various classes of amorphous materials.

3. *Experimentalists should realize that not all of the highly rewarding measurements on crystalline materials are necessarily fruitful when applied to amorphous semiconductors.* Efforts should be concentrated on the acquisition of high-quality physical data (transport, optical, magnetic, etc.) that challenge theoretical interpretation.

4. *Coordinated experimental programs should be encouraged where all aspects of preparation, characterization, and measurement are addressed, results correlated, and cause and effect relationships pursued and understood.*

Preparation, Characterization, and Sample Reproducibility

1. *It is essential to develop a more widespread awareness of the methods of material preparation and characterization, since the structure and properties of an amorphous material may depend intrinsically on the method of preparation.* It is commonly believed that simplicity of sample preparation is an advantage of these materials as compared to crystals. While certain aspects of sample preparation may be simpler, there are also many special problems peculiar to amorphous semiconductors. Presently available characterization tools should be applied more broadly, and new tools should be developed that are useful, especially in the microscopic structure region where diffraction techniques are now inadequate.

2. *Publications should include details concerning preparation, characterization, and measurement sufficient that both samples and measurements on them can be reproduced at other laboratories.* In particular, items known or suspected to affect property measurements directly and estimates of purity and impurity effects should be delineated.

3. *Efforts should always be made to distinguish between experimental properties that are characteristic of the actual materials and artifacts,*

perhaps resulting from poor material preparation. It is presently impossible to specify the actual structure of a given sample completely. The following suggestions may help to overcome these difficulties:

(a) One should try to distinguish between properties that are largely structure sensitive and those that are largely structure insensitive. Useful information will emerge from investigations comparing bulk-quenched and thin-film samples of the same material, and examining the effects of various sample treatments such as quenching and annealing.

(b) The reliability of experimental observations should be ascertained by

i) performing several kinds of experiments—for example, transport and optical—on the same sample

ii) studying the same effect for a group of homologous materials in order to ascertain whether the observed trends are physically reasonable.

(c) Samples may be, and probably are, macroscopically inhomogeneous in many cases. The possible existence of voids and phase separated regions should be recognized and examined experimentally. Inhomogeneities may influence the experimental results and their interpretation profoundly.

4. *It is important to encourage an exchange of samples among different laboratories* to permit identical measurements using different apparatuses as another check on reproducibility. Such "round robin" measurements and subsequent discussions among the investigators involved would serve to clarify questions concerning preparation difficulties and to pinpoint the sources from which disagreements on measurements might arise. For example, ambiguities involving very small Hall voltages, to the extent they exist, might well be clarified in this way.

5. *A better understanding of the dynamics of the preparative process is needed.* Areas receiving very little attention to date, but of potentially great significance, are the vaporization, vapor transport, and condensation processes so frequently used in sample preparation. Solid–liquid–vapor equilibria data, deviations from equilibrium introduced by various low- and high-energy evaporation and sputtering methods, and energy losses during and after condensation could yield significant information about the glass structure and structural variations among samples. Special attention should be given to simple monatomic and binary systems representative of chain, layer, and network structures, and the comparison of bulk-quenched with thin-film samples.

ATOMIC AND ELECTRONIC STRUCTURE

1. *The Committee recommends that investigations directed toward better understanding of how to control the evolution of structure in amorphous semiconductors, e.g., by crystallization or phase separation, be strongly encouraged.* Two of the most promising categories of study in this area are

(a) catalysis of crystallization by photons, trace impurities, or electric fields,
(b) thermodynamics, kinetics, and morphology of phase separation in chalcogenides.

This recommendation is principally motivated by the judgment that some of the most promising applications of amorphous semiconductors result from the ease with which changes in the spatial distribution of phases with different properties can be produced.

2. *Studies of the generation, manually or by computers, of random network models and their dynamic evolution, should be extended, especially to 3-coordinated systems and to systems of two or more components in which the coordination is mixed.* Where possible, the methods of generating these structures ought to be designed to simulate conditions that might obtain in actual deposition processes. Further, the annealing processes following the generation of the structure should be followed by molecular dynamic studies. These studies can provide important insights into the atomic and electronic properties of the ideal structures as well as the occurrence and properties of defects in the actual amorphous structures.

3. *Studies of those aspects of the theory of solid-state cohesion that are especially pertinent to the amorphous state should be extended.* Important aspects are the effects of distortions from ideal bond angles and of dispersion of bond lengths on the cohesive energy and on the electronic structure of the system.

4. *There is a need for good measurements of the temperature dependence of the specific enthalpy and volume of amorphous semiconductors, especially chalcogenides.* Such data, from which the entropy can be calculated, are often crucial for the testing of models for the amorphous state. While available for some molecular and oxide glass formers, they are quite sparse for chalcogenides.

5. *Careful diffraction and spectroscopic studies in conjunction with precise density measurements can greatly delimit the acceptable models for the structure of an amorphous system.* Research on structure de-

termination, using the many excellent spectroscopic and diffraction techniques now available, should be continued. However, the Committee recommends that more emphasis be placed on

(a) techniques for minimizing Compton-modified scattering in x-ray studies,

(b) neutron scattering studies, and

(c) application of scanning electron diffraction techniques to studies of the structure of thin amorphous films.

RADIATION HARDNESS AND RADIATION-INDUCED DEFECTS

One of the unique properties of amorphous semiconductor devices is their low sensitivity to high-energy radiation. There is relatively little in the published literature pointing to extensive investigations of this effect. In view of this, *further and more complete investigations to explore radiation hardness and the reasons for its existence will be of interest.* Tests should be made not only on different devices and geometries, but also on the amorphous materials themselves. From a more fundamental point of view, such investigations should also answer questions about the nature of radiation-induced defects and the electronic states produced by them.

DEVICE-ORIENTED PHYSICS

1. *The Committee underlines the importance of continuing and initiating research aimed at the technological exploitation of the unique properties of amorphous semiconductors.* While the semiconducting and photoconducting properties of these materials will continue to be a source of application possibilities, the ease with which phase changes can be induced through a variety of influences should particularly motivate invention and exploration. Recent demonstrations of applications based on light-induced phase changes (crystallization and phase separation) suggest that this is the beginning of a field rich in possibilities.

2. *In view of the continuing uncertainty about the relative role of electronic and thermal switching mechanisms, it is important to provide information that can distinguish between them.* Although there is still great uncertainty in the formulation of an electronic switching theory, the precise consequence of a purely thermal theory can be worked out at the present time. Such a calculation must take into account the

specific device geometry and the nonlinear electrical conductivity of the device material. More detailed experimental data on the dependence of device behavior on thickness, temperature, deposition parameters, composition, and time sequence of electric fields are also needed.

LEVEL, SCOPE, AND SUPPORT

In recent years, the influence of N. F. Mott and S. R. Ovshinsky and his collaborators has had a particularly strong catalytic effect on the interest in and the development of the area of amorphous semiconductors in this country. Some indication of the present level of research activity in this field and its growth is provided by an examination of contributed papers presented at the Fourth International Conference on Amorphous Semiconductors (Ann Arbor, 1971) and a comparison with previous conferences. Of 126 papers, roughly a third originated abroad. These represent work performed at 40 laboratories in the United States and 31 laboratories in 13 foreign countries. Of the U.S. papers, 40 percent, 50 percent, and 10 percent originated respectively in industrial, academic, and government laboratories. About half of the U.S. work was supported by government funds. In terms of numbers of submitted papers, the 1971 Conference exhibits a substantial increase over the size of the preceding Conference held in 1969, and thus reflects a marked increase in activity. The subject matter of these papers encompasses a broad range extending from the sort of fundamental work described in Sections II and V of this report, to technologically motivated research, such as that typified by Section VI. These international conferences have been most useful in bringing together active contributors and in summarizing the state of research.

Considering the broad range of present and future applications of amorphous materials (from windows and coatings to electronic, optical, and ultrasonic devices), *the Committee urges that a sizable effort of high-quality research regarding the physical, chemical, and engineering properties of amorphous materials be encouraged. The present level of financial support should be maintained and probably increased somewhat, in order to insure a continuing growth of the field.* In connection with the more fundamentally oriented activities, it seems particularly important to encourage research dealing with reproducible experiments on well-characterized materials and to discourage work concerned with uncritical and often ill-considered isolated measurements of one or two effects in poorly prepared materials. Such measurements are unlikely to be reproducible, and their report in the literature may well be counterproductive. While some excellent and high-quality work is being carried on at many laboratories, there is unfortunately still too much information that is produced and accepted by the scientific community without proper critical appraisal. The field of amorphous materials is certainly far more complex than the physics of crystalline materials, but the standard of excellence characterizing research in the latter field should be regarded as a hopefully attainable ideal.

While there have been substantial efforts in some established areas, such as materials for photoconductive applications, the total level of industrial support for the remainder of the field is by itself too small to ensure continuing development. Many companies are conducting small monitoring operations which, to be sure, are frequently of high quality. Except for ECD, few seem to have a major commitment to this area at the present time. Since the field of amorphous semiconductors has a technological potential whose impact cannot yet be accurately predicted, support by governmental agencies should be continued. This could be effectively spread over government, industrial, and university laboratories. For example, more fundamental research of the type that can be profitably pursued at universities is needed. On the other hand, the more technological aspects might well be investigated more intensively in industrial or in-house government laboratories.

Optimal support of research on amorphous materials will lead to a better understanding of the disordered state as well as a clearer assessment of the technological potential of devices incorporating amorphous materials.

References

The convention for references is the following: Each paper is identified by giving the first two letters of the author's name and the year of publication. When necessary, as many additional letters are provided in order to resolve remaining ambiguities (e.g., Mo = Mott, Mor = Morgan). When several papers involving the same first author have appeared in one year, they are distinguished by lower case letters (e.g., 70a, 70b, etc.). Preprints for which the year of publication is not yet known are labeled by a dash following the author code (e.g., Hei —). Review or more comprehensive articles are labeled (R).

AC 70	R	Special review issue of Analytical Chemistry published in April of each even year. V42, 1970 is the latest.
Ad 70	R	D. Adler, Electronics *43*, 61 (September 28, 1970).
Ad 71a		D. Adler and J. Feinleib, in *Physics of Optoelectronic Materials*, W. A. Alvers, ed. (Plenum, New York, 1971).
Ad 71b	R	D. Adler, International Journal of Magnetism, 1971 (to be published).
Ai 69		R. N. Aiyer, R. J. Elliott, J. A. Krumhansl, and P. L. Leath, Phys. Rev. *181*, 1006 (1969).
Al 71		G. S. Almasi, 1971 Digests of the Intermag Conference, Denver, April 1971, Paper 2.9.
Am 69		E. A. Amrheim, Phys. Lett. 29A, 329 (1969).

96   FUNDAMENTALS OF AMORPHOUS SEMICONDUCTORS

Amb 71 | V. Ambegaokar, B. I. Halperin, and J. S. Langer, 4th International Conf. on Amorphous and Liquid Semiconductors, Ann Arbor, Mich., August 8-13, 1971 (to be published).

Ame 70 | I. Ames, F. M. d'Heurle, and R. E. Horstmann, IBM J. Res. Develop. 14, 461 (1970).

Ami 65 | A. Amick in Photoelectronic Materials and Devices, S. Larch, ed. (Van Nostrand, Princeton, New Jersey, 1965).

An 58 | P. W. Anderson, Phys. Rev. 109, 1492 (1958).

An 67 | P. W. Anderson and W. L. McMillan, in Proceedings of the International School of Physics "Enrico Fermi", Course 37, edited by W. Marshall (Academic Press, New York 1967) p. 50.

An 70 | P. W. Anderson, Comments on Sol. State Phys. 2, 193 (1970).

Ar 68 | F. Argall and A. K. Jonscher, Thin Sol. Films 2, 185 (1968).

Au 69   R | I. G. Austin and N. F. Mott, Adv. in Phys. 18, 41 (1969).

Auf 70 | R. Auf der Heide, Datamation, July 15, 1970, p. 66.

Ba 64 | L. Banyai, Physique des Semiconducteurs (Paris, Dunod, 1964) p. 417.

Bag 70 | B. G. Bagley, Sol. State Comm. 8, 345 (1970).

Bal 70 | I. Balberg, Appl. Phys. Letters 16, 491 (1970).

Bas 71 | S. Basavaiah and K. C. Park, IEEE Trans. on Electron Devices, 1971 (to be published).

Be 59 | J. D. Bernal, Nature 183, 141 (1959).

Be 60a | J. D. Bernal, Nature 185, 68 (1960).

Be 60b | J. D. Bernal and J. Mason, Nature 188, 910 (1960).

Bel 66 | R. J. Bell and P. Dean, Nature 212, 1354 (1966).

Bel 68 | R. J. Bell and P. Dean, Phys. Chem. Glasses 9, 125 (1968).

Ben 70 | C. H. Bennett, Ph.D. Thesis, Dept. of Chemistry, Harvard University, Cambridge, Massachusetts 02138 (1970) also to be published.

Bet 70 | F. Betts, A. Bienenstock, and S. R. Ovshinsky, J. Non-Crystalline Solids 4, 554 (1970).

Bi 70 | A. Bienenstock, F. Betts, and S. R. Ovshinsky, J. Non-Crystalline Solids 2, 347 (1970).

Bo 69 | Z. U. Borisova, A. V. Pazin and E. A. Egorova, Russ. J. Appl. Chem. 42, 1988 (1969).

Böe 70a | K. W. Böer and R. Haislip, Phys. Rev. Letters 24, 230 (1970).

Böe 70b | K. W. Böer, Phys. Stat. Sol. 2, 817 (1970).

Böe 71 | K. W. Böer, Phys. Stat. Sol. 4, 571 (1971).

Bon 70 | P. I. Bonyhard, I. Danylchuk, D. E. Kish and J. L. Smith, IEEE Transactions on Magnetics, MAG-6, 447 (1970).

Bonc 62 | V. L. Bonch-Bruevich, in Proc. Int. Conf. on Semiconductor Physics, Exeter, 1962 (Inst. of Phys. and the Phys. Soc., London, 1963) p. 216.

Bor 63 | R. E. Borland, Proc. Roy. Soc. (London) A 274, 529 (1963).

Bos 70 | J. R. Bosnell and U. C. Voisey, Thin Solid Films 6, 161 (1970).

Bosm 66 | A. J. Bosman and C. Crevecoeur, Phys. Rev. 144, 763 (1966).

Boy 71 | A. J. Boyle and J. McNichol, The Electronic Engineer, Feb. 1971, p. 38.

Br 69 | M. H. Brodsky and R. S. Title, Phys. Rev. Lett. 23, 581 (1969).

Br 70   M. H. Brodsky and P. J. Stiles, Phys. Rev. Lett. *25*, 798 (1970).

Br 71a   M. H. Brodsky and D. Turnbull, Bull. Am. Phys. Soc. *16*, 304 (1971).

Br 71b   R   M. H. Brodsky, J. Vac. Sci. and Tech. *8*, 125 (1971).

Bra 70   R. G. Brandes, F. P. Laming, and A. D. Pearson, Applied Optics *9*, 1712 (1970).

Bre 69   G. Breitling and H. Richter, Mat. Res. Bull. *4*, 19 (1969).

Bri 29   G. Brieglieb, Z. Physik. Chem. *A144*, 321 (1929).

Brin 70   W. F. Brinkman and T. M. Rice, Phys. Rev. *B2*, 1324 (1970).

Bro 60   J. J. Brophy, IRE Trans. on Audio, *AU-8*, 58 (1960).

Bru 69   D. Brust, Phys. Rev. Letters *23*, 1232 (1969), Phys. Rev. *186*, 768 (1969).

Bu 71   R   R. H. Bube, ed. *Electronic Properties of Materials*, McGraw-Hill Book Company, N.Y. to be published 1971.

Bus 70   G. Busch, H. J. Güntherodt, H. U. Künzi, and A. Schweiger, Phys. Letters *33A*, 64 (1970).

Ca 65   J. W. Cahn, J. Chem. Phys. *42*, 93 (1965).

Ca 68   R   J. W. Cahn, Trans. Met. Soc. A.I.M.E. *242*, 166 (1968).

Cam 70   R   W. J. Campbell and J. V. Gilfrich, Anal. Chem. *42*, 248 (1970).

Car 70   G. S. Cargill, III, J. Appl. Phys. *41*, 2248 (1970).

Ch 69   R   K. L. Chopra, *Thin Film Phenomena* McGraw-Hill Book Company, N.Y., 1969.

Ch 70   K. L. Chopra and S. K. Bahl, Phys. Rev. *B1*, 2545 (1970).

Cha 69   W. S. Chan and A. K. Jonscher, Phys. Stat. Sol. *32*, 749 (1969).

Chap 69   R. Chapman, cited by Product Engineering *40*, 95 (Sept. 22, 1969).

Chapm 70   D. W. Chapman, Proc. of the IEEE Computer Group Conference, Washington, D.C., June 1970, p. 56.

Chi 67   Y. S. Chiang and K. Johnson, J. Appl. Phys. *38*, 1647 (1967).

Ci 70   Z. Cimpl, F. Kosek and M. Matyiv, Phys. Stat. Sol. *41*, 535 (1970).

Cl 67   A. H. Clark, Phys. Rev., *154*, 750 (1967).

Cl 70   A. H. Clark, J. Non-Crystalline Solids *2*, 52 (1970).

Co 71   G. A. N. Connell and W. Paul, Bull. Am. Phys. Soc. *16*, 347 (1971).

Coc 70   R   G. C. Cocks, Anal. Chem. *42*, 114R (1970).

Cog 71   Product Literature, *Monolithic Memory Systems Summary Catalog*, Cogar Corporation, 1971.

Coh 60   M. H. Cohen and D. Turnbull, J. Chem. Phys. *34*, 120 (1960).

Coh 64   M. H. Cohen and D. Turnbull, Nature *203*, 964 (1964).

Coh 69   M. H. Cohen, H. Fritzsche, and S. R. Ovshinsky, Phys. Rev. Letters *22*, 1065 (1969).

Coh 71a   M. H. Cohen and K. Freed, Phys. Rev., 1971 (to be published).

Coh 71b   M. H. Cohen, Physics Today *24*, 26 (1971).

Com 71   Computer *4*, March-April, 1971.

Con 54   R   E. U. Condon, Am. J. Phys. *22*, 43 (1954).

Cr 70   R   R. O. Crisler, Anal. Chem. *42*, 388R (1970).

Cs 70   A. Csillag and H. Jäger, J. Non-Crystalline Solids *2*, 133 (1970).

Cy 66   F. Cyrot-Lackmann, J. Phys. (France) *27*, 627 (1966).

Da 70   E. A. Davis and N. F. Mott, Phil. Mag. *22*, 903 (1970).
Dam 68   S. Damron and E. K. Kietz, Modern Data, Dec. 1968, p. 28.
De 69   S. A. Dembovsky, Phys. and Chem. of Glasses *10*, 73 (1969).
Dea 71   R G. Dearnaley, A. M. Stoneham, and D. V. Morgan, "Electrical Phenomena in Amorphous Oxide Films," to appear in Reports on Progress in Physics.
DeN 71   J. P. DeNeufville and D. Turnbull, "The Vitreous State," Discussions of the Faraday Soc. to be published (1971). See also: J. P. DeNeufville, Ph.D. thesis, Harvard University, Cambridge, Mass. (1969).
Des 65   R J. H. Dessauer and H. E. Clark eds. *Xerography and Related Processes* (Focal Press, London and New York, 1965).
Dew 62   J. F. Dewald, A. D. Pearson, W. R. Northover, and W. F. Peck, J. Electrochem. Soc. *109*, 243C (1962).
Do 70   T. M. Donovan, W. E. Spicer, J. M. Bennett and E. J. Ashley, Phys. Rev. *B2*, 397 (1970).
Dow 70   J. D. Dow and D. Redfield, Phys. Rev. *B1*, 3358 (1970). Phys. Rev. Letters *26*, 762 (1971).
Dr 68   J. Dresner and G. B. Stringfellow, J. Phys. Chem. Solids *29*, 303 (1968).
Dra 69   C. F. Drake, I. F. Scanlan, and A. Engel, Phys. Stat. Sol. *32*, 193 (1969).
Du 70   P. Duwez, "Bibliography on Alloys Quenched from the Liquid State" publication CALT-822-17, of the California Institute of Technology, December 1970.
Dz 67   S. U. Dzhalilov and K. I. Rzaev, Phys. Stat. Sol. *20*, 261 (1967).
Ec 70a   E. N. Economou, S. Kirkpatrick, M. H. Cohen, and T. P. Eggarter, Phys. Rev. Letters *25*, 520 (1970).
Ec 70b   E. N. Economou, and M. H. Cohen, Phys. Rev. Letters *25*, 1445 (1970).
ECD   ECD Product Literature "Preliminary Specifications, Ovonic Read-Mostly Memory RM-256."
Ed 67   S. F. Edwards, Adv. Phys. *16*, 147 (1967).
Edw 71   J. T. Edwards and D. J. Thouless, J. Phys. C *4*, 1971 (to be published).
Ef 69   A. Efstathiou, D. M. Hoffman, and E. R. Levin, J. Vac. Sci. and Tech. *6*, 383 (1969).
Eg 70   T. P. Eggarter and M. H. Cohen, Phys. Rev. Letters *25*, 807 (1970).
Eh 70   R H. Ehrenreich and D. Turnbull, Comments on Solid State Physics *3*, 75 (1970).
El 71a   Electronic News, Feb. 22, 1971, p. 39.
El 71b   Electronic News, March 29, 1971, p. 8.
El 71c   Electronic News, May 3, 1971, p. 33.
En 70   W. A. England, IEEE Trans. on Magnetics, *MAG-6*, 528 (1970).
Eng 70   W. E. Engeler, J. J. Tiemann, and R. D. Baertsch, Appl. Phys. Letters *17*, 469 (1970).
Es 70   L. Esaki and L. L. Chang, Phys. Rev. Letters *25*, 653 (1970).
Esc 70   A. H. Eschenfelder, J. Appl. Phys. *41*, 1372 (1970).
Ev 66   D. L. Evans and S. R. King, Nature *212*, 1353 (1966).

Eva 68   E. J. Evans, "A Feasibility Study of the Application of Amorphous Semiconductors to Radiation Hardening of Electronic Systems," Picatinny Arsenal Technical Report 3698, July 1968.

Evan 70   R   C. A. Evans, Jr. and J. P. Remsler, Anal. Chem. *42*, 1060 (1970).

Fa 70   E. A. Fagen and H. Fritzsche, J. Non-Crystalline Solids *2*, 170 (1970).

Fan 71   G. J. Fan, 1971 Digests of the Intermag. Conference, Denver, April 1971, Paper 19.1.

Fe 71   R   R. Fernquist and J. Short, *Proc. 14th Annual Technical Conference of the Society of Vacuum Coaters*, 1971 (to be published).

Fei 71a   J. Feinleib, J. DeNeufville, S. C. Moss, and S. R. Ovshinsky, Appl. Phys. Letters *18*, 254 (1971).

Fei 71b   J. Feinlcib, S. Iwasa, S. C. Moss, J. DeNeufville, and S. R. Ovshinsky, *Fourth International Conference on Amorphous and Liquid Semiconductors*, Ann Arbor, Michigan, August 8-13, 1971 (to be published).

Fi 71a   J. E. Fischer, 4th International Conf. on Amorphous and Liquid Semiconductors, Ann Arbor, Mich., August 8-13, 1971 (to be published).

Fi 71b   J. E. Fischer and T. M. Donovan, Optics Communications *3*, 116 (1971).

Fis 71   R. Fischer, U. Heim, F. Stern, K. Weiser, Phys. Rev. Lett. *26*, 1182 (1971).

Fisc 71   R. F. Fischer, 1971 Digests of the Intermag. Conference, Denver, April 1971, Paper 28.2.

Fish 70   R   R. M. Fisher, A. Szirmae, and J. M. McAleav, Anal. Chem. *42*, 362R (1970).

Fl 70   T. M. Flanagan and M. E. Wyatt, J. Non-Crystalline Solids *2*, 229 (1970).

Fla 60   S. S. Flaschen, A. D. Pearson, and I. L. Kalnins, J. Appl. Phys. *31*, 431 (1960).

Fr 70a   H. Fritzsche and S. R. Ovshinsky, J. Non-Crystalline Solids *2*, 393 (1970).

Fr 70b   H. Fritzsche and S. R. Ovshinsky, J. Non-Crystalline Solids *4*, 464 (1970).

Fr 71a   R   H. Fritzsche, Chapter 13 in *Electronic Properties of Materials*, edited by R. H. Bube (1971).

Fr 71b   R   H. Fritzsche, J. Non-Crystalline Solids, 1971 (to be published).

Fra 56   W. Franz, Handbuch der Physik *17, 155*, S. Flugge, ed. (Springer, Berlin, 1956).

Fre 69   P. J. Freud and A. Z. Hed, Phys. Rev. Letters *23*, 1440 (1969).

Fro 70   D. Frohman-Bentchkowsky, Proc. of IEEE *58*, 1207 (1970).

Gi 58   J. Gibbs and E. A. DiMarzio, J. Chem. Phys. *28*, 373 (1958).

Gil 70   J. C. G. Gilbert, Wireless World *76*, 377 (1970). Supplemented by Teldec press release material.

Gr 67   R. Grigorovici, N. Croitoru, and A. Devenyi, Phys. Status Solidi *23*, 621 (1967).

Gr 68   R. Grigorovici, Mat. Res. Bull. *3*, 13 (1968).

Gr 69a   R. Grigorovici, J. Non-Crystalline Solids *1*, 303 (1969).

Gr 69b   R. Grigorovici and R. Manaila, J. Non-Crystalline Solids *1*, 371 (1969).

Gre 64   I. Greenberg, IEEE Spectrum, Nov. 1964, p. 75.

Gu 63   A. I. Gubanov, *Quantum Electron Theory of Amorphous Semiconductors* (1963). (Translated by Consultants Bureau, N.Y., 1965.)

Ha 65   B. I. Halperin, Ph.D. Thesis, Univ. of Calif., Berkeley, 1965 (Available from Univ. Microfilms, Ann Arbor, Mich.).

Ha 66   B. I. Halperin and M. Lax, Phys. Rev. *148*, 722 (1966), *153*, 802 (1967).

Ha 68   B. I. Halperin, Advances in Chemical Physics *13*, 123 (1968).

Hab 70   D. R. Haberland, Solid-State Electronics *13*, 207 (1970).

Hal 65   W. Haller, J. Chem. Phys. *42*, 686 (1965).

Har 70   W. J. Harmon, Jr., 1970 IEEE International Convention Digest, New York, March 23-26, 1970, p. 72.

Harr 68   R   L. A. Harris, Anal. Chem. *40* (14), 24A (1968).

He 70   R   D. M. Hercules, Anal. Chem. *42* (1), 20A (1970). See also D. Betteridge and A. D. Baker, Anal. Chem. *42* (1), 43A (1970).

Hee 70   R   J. P. Heeschen, Anal. Chem. *42*, 418R (1970).

Hei 69   V. Heine and R. O. Jones, J. Phys. C *2*, 719 (1969).

Hei   V. Heine (to be published).

Hel 71   J. H. Helbers, "Read Mostly Memory Using Ovonic/Bipolar Arrays," written version of a paper presented at the 1971 Computer Designer's Conference, Anaheim, California, January 19-21, 1971.

Hen 71   D. Henderson, Bull. Am. Phys. Soc. *16*, 348 (1971).

Henc 70   L. L. Hench, J. Non-Crystalline Solids *2*, 250 (1970).

Heni 69   H. K. Henisch, Scientific American *221*, 30 (1969).

Heni 70   H. K. Henisch, E. A. Fagen, and S. R. Ovshinsky, J. Non-Crystalline Solids *4*, 538 (1970).

Heni 71   H. K. Henisch and R. W. Pryor, Solid State Electronics, 1971 (to be published).

Her 60   C. Herring, J. Appl. Phys. *31*, 1939 (1960).

Herr 71   D. J. Herrell and K. C. Park, 4th International Conference on Amorphous and Liquid Semiconductors, Ann Arbor, Mich., August 8-13, 1971 (to be published).

Hi 66   A. R. Hilton and C. E. Jones, Phys. and Chem. of Glasses *7*, 112 (1966).

Hil 68   R   J. E. Hilliard, "Phase Transformations" pp. 497-560, A.S.M. Seminar volume (1968) Metals Park, Ohio.

Ho 70   W. E. Howard and R. Tsu, *Proceedings of the 10th International Conference on the Physics of Semiconductors*, Cambridge, Mass., 1970, p. 789.

Hom 71   K. Homma, Appl. Phys. Letters *18*, 198 (1971).

Hop 61   J. J. Hopfield, J. Phys. Chem. Sol. *22*, 63 (1961).

Hop 68   J. J. Hopfield, Comments on Solid State Physics, *1*, 16 (1968).

Hopn 69   E. Hopner, AGEN, Nr. 10, p. 48, Sept. 1969.

Hor 68   J. Hori, *Spectral Properties of Disordered Chains and Lattices* (Pergamon Press, Oxford, 1968).

Hov 70 — H. J. Hovel, Appl. Phys. Letters *17*, 141 (1970).

Hov 71 — H. J. Hovel, J. Appl. Phys., 1971 (to be published).

Hu 70 — R. P. Hunt, T. Elser, and I. W. Wolf, Datamation, April 1970, p. 97.

IE 71 — Digest, IEEE 1971 International Convention, March 22-25, 1971, New York, N.Y. Several relevant papers appear in this volume.

In— — S. W. Ing, J. Neyhart, and F. Schmidlin, J. Appl. Phys. (to be published).

In 69 — S. W. Ing, Y. S. Chiang and A. Ward, Communication to the American Ceramic Society Meeting (May 1969).

Io 60 — A. F. Ioffe and A. R. Regel, Progr. Semiconductors *4*, 239 (1960).

Iv 63 — H. F. Ivey, *Electroluminescence and Related Effects* (Academic Press, New York, 1963), p. 74.

Jo 71 — A. K. Jonscher, J. of Vac. Sci. and Tech. *8*, 135 (1971).

Joh 70 — R. L. Johnson, "The Application of the Plasma Display Technique to Computer Memory Systems," Coordinated Science Laboratory Report R-461, University of Illinois, April 1970.

Ka 62 — E. O. Kane, Phys. Rev. *125*, 1094 (1962); *131*, 1532 (1963).

Ka 63 — E. O. Kane, Phys. Rev. *131*, 79 (1963).

Kau 48 R — W. Kauzman, Chem. Rev. *43*, 219 (1948).

Kaz 68 — B. Kazan and M. Knoll, *Electronic Image Storage* (Academic Press, New York, 1968).

Ke 67 — R. C. Keezer and M. W. Bailey, Mat. Res. Bull. *2*, 185 (1967).

Kei 66 — T. H. Keil, Phys. Rev. *144*, 582 (1966).

Ken 70a — S. A. Keneman, G. W. Taylor, A. Miller, and W. H. Fonger, Appl. Phys. Letters *17*, 173 (1970).

Ken 70b — S. A. Keneman, G. W. Taylor, and A. Miller, Ferroelectrics *1*, 227 (1970).

Key 71 — R. W. Keyes, "Thermal Negative Resistance in Amorphous Semiconductors," to be published in Comments on Solid State Physics (1971).

Ki 70 — S. Kirkpatrick, B. Velický and H. Ehrenreich, Phys. Rev. *B1*, (1970).

Kit 66 — C. Kittel, *Introduction to Solid State Physics* (Wiley, New York, 1966) 3rd ed.

Kl 71 — P. Klose, 4th International Conf. on Amorphous and Liquid Semiconductors, Ann Arbor, Mich., August 8-13, 1971 (to be published).

Kla 61 — J. R. Klauder, Ann. Phys. (N.Y.) *14*, 43 (1961).

Ko 60 — B. T. Kolomiets and T. F. Nazarova, Sov. Phys.—Solid State *2*, 369 (1960).

Ko 62 R — B. T. Kolomiets, T. F. Nazarova and V. P. Shilo, *Proc. Intern. Conf. Phys. Semicond.*, Exeter, England 1962, 259.

Ko 64 — B. T. Kolomiets, Phys. Stat. Sol. *7*, 359 (1964).

Ko 69 — B. T. Kolomiets, E. A. Lebedev, and I. A. Taksami, Soviet Phys. Semiconductors *3*, 267 (1969).

Ko 70 — B. T. Kolomiets, B. T. Mamontova and A. A. Babaev, J. Non-Crystalline Solids *4*, 289 (1970).

102   FUNDAMENTALS OF AMORPHOUS SEMICONDUCTORS

Kon 70   D. C. Koningsberger and T. DeNeef, Chem. Phys. Letters *4*, 615 (1970).

Kot 67   B. A. Kotov, N. M. Okuneva, A. R. Regel and A. L. Shakh-Budagov, Sov. Physics-Solid State *9*, 955 (1967).

Kr 69   R   H. Krebs, J. Non-Crystalline Solids *1*, 455 (1969).

Kra 70   J. T. Krause, C. R. Kurkjian, D. A. Pinnow and E. A. Sigeby, Appl. Phys. Lett. *17*, 367 (1970).

Kram 70   B. Kramer, Phys. Rev. Letters *25*, 1020 (1970), Phys. Stat. Sol. *41*, 649, 725 (1970).

La 70   W. C. LaCourse, V. A. Twaddell and J. D. Mackenzie, J. Non-Crystalline Solids *3*, 234 (1970).

Lan 70   R. Landauer, Phil. Mag. *21*, 863 (1970).

Lax 51   M. Lax, Rev. Mod. Phys. *23*, 287 (1951); Phys. Rev. *85*, 621 (1952).

Lee 71   G. H. Lee and H. K. Henisch, 4th International Conf. on Amorphous and Liquid Semiconductors, Ann Arbor, Mich., August 8-13, 1971 (to be published).

Li 63   I. M. Lifshitz, Zh. Eksperim i Teor. Fiz. *44*, 1723 (1963) (Sov. Phys. JETP *17*, 1159 (1963)). See also Adv. Phys. *13*, 483 (1964); Nuovo Cimento Suppl. *3*, 716 (1956); Usp. Fiz. Nauk. *83*, 617 (1964); (Sov. Phys. Usp. *7*, 549 (1965)).

Ll 69   P. Lloyd, J. Phys. C *2*, 1717 (1969).

Lu 69   G. Lucovsky, Mat. Res. Bull. *4*, 505 (1969).

Lu 70   G. Lucovsky, *Proceedings of the 10th International Conference on the Physics of Semiconductors*, Cambridge, Mass., 1970, p. 799.

Lue 35   H. Lueder and E. Spenke, Phys. Zeitschr. *36*, 767 (1935).

Ma 64   C. H. Massen, A. G. L. M. Weijts and J. A. Poulis, Trans. Faraday Soc. *60*, 317 (1964).

MAB 67   "Characterization of Materials" Publication MAB-229-M of the Materials Advisory Board, NAS-NAE, March 1967.

MAB 68   "Infrared Transmitting Materials," Publication MAB-243, of the Materials Advisory Board, NAS-NAE, July 1968.

MAG 70   A series of papers on magneto-optics and thermomagnetics, IEEE Trans. on Magnetics, *MAG-6*, p. 537-572 (1970).

Mai 70   R   L. I. Maissel and R. Glang, ed. *Handbook of Thin Film Technology*, McGraw-Hill Book Company, N.Y. 1970.

Mal 67   J. C. Male, Brit. J. Appl. Phys. *18*, 1543 (1967).

Mat 71   R   R. E. Matick, "Review of Current Proposed Technologies for Mass Storage Systems," 1971 (to be published).

Mau 64   R. D. Maurer, "Symposium on Nucleation and Crystallization in Glasses and Melts" (Edited by Resev and Smith) pp. 5-9 Am. Ceramic Soc., Columbus, Ohio (1964).

Mc 70   T. C. McGill and J. Klima, J. Phys. C *3*, L163 (1970).

McC 70   J. P. McCallister, IEEE Trans. on Magnetics, *MAG-6*, 525 (1970).

McL 64   D. A. McLean, N. Schwartz, E. D. Tidd, Proc. IEEE *52*, 1450 (1964).

McL 66   D. A. McLean and W. H. Orr, Bell Lab. Record *44*, 305 (1966).

Me 70   G. K. Megla and D. R. Steinberg, Information Display *7*, 31 (1970).

Mi 71		D. L. Mitchell, S. G. Bishop and P. C. Taylor, Bull. APS *16*, 303 (1971).
Mo 61		N. F. Mott and W. D. Twose, Adv. Phys. *10*, 107 (1961).
Mo 67a		N. F. Mott and R. S. Allgaier, Phys. Stat. Sol. *21*, 343 (1967).
Mo 67b		N. F. Mott, Adv. Phys. *16*, 49 (1967).
Mo 68		N. F. Mott, *Notes on Ovonic and Memory Switching Devices*, unpublished memorandum (1968).
Mo 69a	R	N. F. Mott, Phil. Mag. *19*, 835 (1969).
Mo 69b		N. F. Mott, Contemporary Physics *10*, 125 (1969).
Mo 69c	R	N. F. Mott in *Festkörperprobleme IX*, ed. O. Madelung (Pergamon Vieweg, Braunschweig, 1969).
Mon 42		E. W. Montroll, J. Chem. Phys. *10*, 218 (1942); *11*, 481 (1943).
Mor 65		T. N. Morgan, Phys. Rev. *139*, A 343 (1965).
Morr		J. E. Morral, Ph.D. Thesis, Department of Materials Science, M.I.T., 1968 (to be published).
Mort 71		J. Mort and H. Scher, Phys. Rev. *B3*, 334 (1971).
Mos 69		S. C. Moss and J. F. Graczyk, Phys. Rev. Letters *23*, 1167 (1969).
Moz 69		R. F. Mozzi and B. E. Warren, Journal of Applied Crystallography *2*, 164 (1969).
Moz 70		R. F. Mozzi and B. E. Warren, Journal of Applied Crystallography *3*, 251 (1970).
MS 70		Product announcement by Memory Systems, Inc. Electronic Design *18*, 95 (1970).
Mu 70	R	C. B. Murphy, Anal. Chem. *42*, 268R (1970).
Mul 70	R	L. N. Mulay and I. L. Mulay, Anal. Chem. *42*, 325R (1970).
My 67		M. B. Myers and E. J. Felty, Mat. Res. Bull. *2*, 535 (1967).
Ne 63		S. V. Nemilov and G. T. Petrovskii, Russ. J. Appl. Chem. *36*, 932 (1963).
Nea 70a		R. G. Neale, D. L. Nelson and G. E. Moore, Electronics *43*, 56 (Sept. 28, 1970).
Nea 70b		R. G. Neale, J. Non-Crystalline Solids *2*, 558 (1970).
Ni 71		R. V. Nicolades and W. Doremus, 4th International Conf. on Amorphous and Liquid Semiconductors, Ann Arbor, Mich., August 8-13, 1971 (to be published).
No 69		A. S. Nowick, Comments on Solid State Phys. *2*, 155 (1969).
Nor 31		L. Nordheim, Ann. Physik *9*, 607, 641 (1931).
Nw 68		A. Nwachuku and M. Kuhn, Appl. Phys. Letters *12*, 13 (1968).
Od 64		J. J. O'Dwyer, *The Theory of Dielectric Breakdown of Solids* (Oxford University Press, Oxford 1964).
Ov 68a		S. R. Ovshinsky, Phys. Rev. Lett. *21*, 1450 (1968).
Ov 68b		S. R. Ovshinsky, E. J. Evans, D. L. Nelson, and H. Fritzsche, IEEE Trans. Nucl. Sci., *NS-15*, 311 (December 1968).
Ov 71a		S. R. Ovshinsky and H. Fritzsche, Metallurgical Transactions *2*, 641 (1971).
Ov 71b		S. R. Ovshinsky and P. H. Klose, *Digest of Technical Papers*, 1971 SID International Symposium, Philadelphia, May 1971, p. 58 (Lewis Winner, New York, 1971).
Ov 71c		S. R. Ovshinsky, private communication (1971).
Ow 67		A. E. Owen, Glass Ind. *48*, 637, 695 (1967).
Ow 70a		A. E. Owen, Cont. Phys. *11*, No. 3, p. 227 (1970).

Ow 70b   A. E. Owen and J. M. Robertson, J. Non-Crystalline Solids *4*, 40 (1970).

Pat 71   A. M. Patlach, W. E. Schillinger, and B. R. Brown, 1971 Digests of the Intermag. Conference, Denver, April, 1971, Paper 6.9.

Pe 68   I. N. Pen'kov and I. A. Safin, Soviet Physics-Crystallography *13*, 264 (1968).

Pea 70   A. D. Pearson, J. Non-Crystalline Solids *2*, 1 (1970).

Pen 62   D. R. Penn, Phys. Rev. *128*, 2093 (1962).

Pet 70   R. J. Petschauer, Computer, Vol. 3, No. 6, p. 13, Nov.-Dec. 1970.

Pf 70   R   C. E. Pfluger, Anal. Chem. *42*, 317R (1970).

Ph 70a   J. C. Phillips, Rev. Mod. Phys. *42*, 317 (1970).

Ph 70b   J. C. Phillips, Comments on Solid State Physics *3*, 105 (1970).

PI 70   Sales literature on Unicon, available from Precision Instruments, Palo Alto, California.

Po 61   M. Pollak and T. H. Geballe, Phys. Rev. *122*, 1742 (1961).

Pol 71   D. E. Polk, J. of Non-Crystalline Solids *5*, 365 (1971).

PP 70   IBM Product Publication, "Component Summary—3830 Storage Control, 3330 Disk Storage." Order No. GA 26-1592-0, June 1970.

Pr 71   R. W. Pryor and K. Henisch, Appl. Phys. Letters *18*, 324 (1971).

Ra 70   J. A. Rajchman, J. Appl. Phys. *41*, 1376 (1970).

Re 63   D. Redfield, Phys. Rev. *130*, 916 (1963).

Ree 55   A. H. Reeves and R. B. W. Cooke, Electrical Communication *32*, 112 (1955).

Ro 70   R. Roy, J. Non-Crystalline Solids *3*, 33 (1970).

Rom 70   L. T. Romankiw, I. M. Croll, and M. Hatzakis, IEEE Trans. on Magnetics, *MAG-6*, 597 (1970), and following two papers.

Ros 51   A. Rose, RCA Review *12*, 303 (1951), and following set of papers.

Ru 71   W. E. Rudge and I. B. Ortenberger, Bull. Am. Phys. Soc. *16*, 371 (1971).

Sa 68   P. T. Sarjeant and R. Roy, Mat. Res. Bull. *3*, 265 (1968).

San 70   F. L. J. Sangster, 1970 International Solid State Circuits Conference, Univ. of Pennsylvania, February 1970, Digest of Technical Papers, p. 74.

Sau 71   J. A. Sauvage and C. J. Mogab, 4th International Conf. on Amorphous and Liquid Semiconductors, Ann Arbor, Mich., August 8-13, 1971 (to be published).

Sav 65   J. A. Savage and S. Nielsen, Infrared Physics *5*, 195 (1965).

Sc 70   J. Schottmiller, M. Tabak, G. Lucovsky and A. Ward, J. Non-Crystalline Solids *4*, 80 (1970).

Sch 70   M. E. Scharfe, Phys. Rev. *B2*, 5025 (1970).

Scha 65   R   R. M. Schaffert, Electrophotography (Focal Press, London and New York, 1965).

Scha 71   R. M. Schaffert, IBM J. Res. Develop. *15*, 75 (1971).

Schw 71   L. Schwartz and H. Ehrenreich, Annals of Physics *64*, 100 (1971).

Se 68	T. P. Seward, III, D. R. Uhlmann, and D. Turnbull, J. Am. Ceram. Soc. *51*, 634 (1968).
Sh 71	N. G. Shevchik, Bull. Am. Phys. Soc. *16*, 347 (1971).
Sha 70a	R. R. Shanks, J. Non-Crystalline Solids *2*, 504 (1970).
Sha 70b	R. R. Shanks, D. L. Nelson, R. L. Fowler, H. C. Chambers, and D. J. Niehaus, Air Force Avionics Laboratory Technical Report AFAL-TR-70-15, March 1970.
Sha 70c	R. R. Shanks, J. H. Helbers, and D. L. Nelson, Air Force Avionics Laboratory Technical Report AFAL-TR-69-309, June 1970.
Shi 71	L. K. Shick, J. W. Nielsen, A. H. Bobeck, A. J. Kurtzig, P. C. Michaelis, and J. P. Reekstin, Appl. Phys. Letters *18*, 89 (1971).
Si 70	J. G. Simmons, Contemp. Phys. *11*, 21 (1970).
SID 71	*Digest of Technical Papers*, 1971 SID International Symposium, Philadelphia, May 1971 (Lewis Winner, New York, 1971).
Sie 71	C. Sie, P. Dugan, and S. C. Moss, 4th International Conf. on Amorphous and Liquid Semiconductors, Ann Arbor, Mich., August 8-13, 1971 (to be published).
Sm 71	J. E. Smith, M. H. Brodsky, B. L. Crowder, M. I. Nathan and A. Pinczuk, Phys. Rev. Lett. *26*, 642 (1971).
Smi 71	R. A. Smith, R. Sanford, and P. E. Warnock, 4th International Conf. on Amorphous and Liquid Semiconductors, Ann Arbor, Mich., August 8-13, 1971 (to be published).
Smit 71	R. E. Smith, 4th International Conf. on Amorphous and Liquid Semiconductors, Ann Arbor, Mich., August 8-13, 1971 (to be published).
So 66	P. Soven, Phys. Rev. *151*, 539 (1966).
So 67	P. Soven, Phys. Rev. *156*, 809 (1967).
Sp 67	W. E. Spicer, Phys. Rev. *154*, 385 (1967).
Sp 70	W. E. Spicer and T. M. Donovan, Phys. Rev. Lett. *24*, 595 (1970).
Spe 36	E. Spenke, Wiss. Veröff. Siemens-Werke *15*, 92 (1936).
Sr 70	Science Research Council (United Kingdom): "The Physics of Amorphous Materials" (June, 1970).
SS 70	IEEE Journal of Solid State Circuits, October 1970, *SC-5*, Papers from p. 174 to p. 235.
SSC 71	Digest of International Solid State Circuits Conference, Feb. 1971. Univ. of Pennsylvania.
St 69 R	J. Stuke, Festkörperprobleme *9*, 46 (1969).
St 70a R	J. Stuke, J. Non-Crystalline Solids *4*, 1 (1970).
St 70b R	J. Stuke, *Proc. Tenth Int. Conference Physics of Semiconductors*, AEC, 1970, p. 14.
Ste 71	R. B. Stephens, R. C. Zeller and R. O. Pohl, Bull. Am. Phys. Soc. *16*, 377 (1971).
Ster 71a	F. Stern, Phys. Rev. *B3*, 2636 (1971).
Ster 71b	F. Stern, *Proceedings of International Conference on Conduction and Low Mobility Materials*, Eilath, Israel (1971).
Sto 70	H. J. Stocker, C. A. Barlow, Jr., and D. F. Weirauch, J. Non-Crystalline Solids *4*, 523 (1970).
Sy 68	Product Literature on "P-Series Hermetic Sealed Electro-

luminescent (EL) Readout Panels," Sylvania Electronics Components, Electronic Tube Division, Emporium, Pa. July 1968, p. 5.

Ta 65   J. Tauc, Progr. Semiconductors 9, 87 (1965).

Ta 70a   R   J. Tauc, *Optical Properties of Non-Crystalline Solids in the Optical Properties of Solids,* edited by F. Abeles, (North Holland, Amsterdam, 1970), p. 723.

Ta 70b   J. Tauc, A. Menth and D. L. Wood, Phys. Rev. Lett. 25, 749 (1970).

Ta 70c   J. Tauc, A. Abraham, R. Zallen and M. Slade, J. Non-Crystalline Solids 4, 279 (1970).

Ta 70d   J. Tauc, Mat. Res. Bull. 5, 721 (1970).

Ta—   J. Tauc, *Optical Properties of Amorphous Semiconductors in Amorphous and Liquid Semiconductors,* edited by J. Tauc, (Plenum Press, London, to be published).

Tab 68   M. D. Tabak and P. J. Warter, Phys. Rev. 173, 899 (1968).

Tab 71   M. D. Tabak, S. W. Ing and M. E. Scharfe, IEEE Transactions on Electron Devices, 1971 (to be published).

Te 70   L. M. Terman, IEEE Trans. on Magnetics, MAG-6, 584 (1970).

Th 71   M. L. Theye, Mat. Res. Bull. 6, 103 (1971).

Tho 70   D. J. Thouless, J. Phys. C 2, 1230 (1970).

Ti 69   U. Tietze and C. Schenk, *Halbleiter Schaltungstechnik,* (Springer, Berlin, 1969).

To 59   Y. Toyozawa, Progr. Theor. Phys. 22, 455 (1959).

Tr 69   M. P. Trubisky and J. H. Neyhart, Applied Optics, Supplement 3, 59 (1969).

Ts 71   S. Tsuchihashi and U. Kawamoto, J. Non-Crystalline Solids 5, 286 (1971).

Tu 58   D. Turnbull and M. H. Cohen, J. Chem. Phys. 29, 1052 (1958).

Tu 69a   R   D. Turnbull, Contemp. Phys. 10, 473 (1969).

Tu 69b   R   D. Turnbull, "Solidification" A.S.M. Seminar Series, A.S.M. Metals Park, Ohio (1969).

Ut 71   N. Utsumi and M. Wada, Japanese J. Appl. Phys. 10, 79 (1971).

Va 68   A. Vaško, Mat. Res. Bull. 3, 209 (1968).

Va 70   A. Vaško, D. Ležal and I. Srb, J. Non-Crystalline Solids 4, 311 (1970).

Van 70   K. E. Van Landingham, G. R. Fleming, D. L. Nelson, "Bulk Semiconductor Switch," Rome Air Development Center Report RADC-TR-70-146, August 1970, p. 53.

VanR 71   W. van Roosbroeck, Bull. Am. Phys. Soc. 16, 348 (1971).

Ve 69   B. Velický, Phys. Rev. 184, 614 (1969).

Ver 57   D. A. Vermilyea, J. Electrochem. Soc. 104, 542 (1957).

Vo 67   A. F. Volkov and Sh. M. Kogan, Soviet Phys.—JETP 25, 1095 (1967).

Wa 69   A. T. Ward and M. B. Myers, J. Phys. Chem. 73, 1374 (1969).

Wal 68   P. A. Walley, Thin Solid Films 2, 327 (1968).

War 37   B. E. Warren, J. Appl. Phys. 8, 645 (1937).

Warr 70   A. C. Warren and J. C. Male, Electronics Letters 6, 567 (1970).

Wart 69   P. J. Warter, Jr., Applied Optics, Supplement 3, 65 (1969).

We 64   K. Weiser and R. S. Levitt, J. Appl. Phys. *35*, 2431 (1964).

We 70   K. Weiser, R. Fischer, and M. H. Brodsky, *Proceedings of the 10th International Conference on the Physics of Semiconductors,* Cambridge, Mass., 1970, p. 667.

Wea 71   D. Weaire, 1971 (to be published).

Wo 71   D. L. Wood, A. Menth, and J. Tauc, Bull. Am. Phys. Soc. *16*, 348 (1971).

Yo 68   F. Yonezawa, Progr. Theor. Phys. *40*, 734 (1968).

Za 71   R. Zallen, R. E. Drews, R. L. Emerald and M. L. Slade, Phys. Rev. Lett. *26*, 1564 (1971).

Zac 32   W. H. Zachariasen, J. Am. Chem. Soc. *54*, 3841 (1932).

Ze 71   R. C. Zeller and R. O. Pohl, Bull. Am. Phys. Soc. *16*, 377 (1971).

Zi 64   J. Ziman, *Principles of the Theory of Solids* (Cambridge University Press, 1964).

Zi 66   J. Ziman, Proc. Phys. Soc. *88*, 387 (1966).

Zi 69   J. Ziman, J. Phys. C 2, 1230 (1969).

Zit 66   Z. Zittartz and J. S. Langer, Phys. Rev. *148*, 741 (1966).